HIDDEN IN PLAIN SIGHT

Beyond the X-Files

Richard Sauder

Adventures Unlimited Press

Other Books by Richard Sauder:
UNDERGROUND BASES AND TUNNELS
UNDERWATER AND UNDERGROUND BASES
KUNDALINI TALES

Other Books of Interest:

THE LOST CONTINENT OF MU
MEN & GODS IN MONGOLIA
IN SECRET MONGOLIA
THE LOST WORLD OF CHAM
ARK OF GOD
VIMANA
ANCIENT TECHNOLOGY IN PERU & BOLIVIA
THE MYSTERY OF THE OLMECS
PIRATES AND THE LOST TEMPLAR FLEET
TECHNOLOGY OF THE GODS
A HITCHHIKER'S GUIDE TO ARMAGEDDON
LOST CONTINENTS & THE HOLLOW EARTH
ATLANTIS & THE POWER SYSTEM OF THE GODS
THE FANTASTIC INVENTIONS OF NIKOLA TESLA
LOST CITIES OF NORTH & CENTRAL AMERICA
LOST CITIES OF CHINA, CENTRAL ASIA & INDIA
LOST CITIES & ANCIENT MYSTERIES OF AFRICA & ARABIA
LOST CITIES & ANCIENT MYSTERIES OF SOUTH AMERICA
LOST CITIES OF ANCIENT LEMURIA & THE PACIFIC
LOST CITIES OF ATLANTIS, ANCIENT EUROPE & THE MEDITERRANEAN
LOST CITIES & ANCIENT MYSTERIES OF THE SOUTHWEST
THE ENIGMA OF CRANIAL DEFORMATION
THE CRYSTAL SKULLS

HIDDEN IN PLAIN SIGHT

Richard Sauder

Hidden In Plain Sight

By Richard Sauder

ISBN: 978-1-948803-34-2

Published by Adventures Unlimited Press
One Adventure Place
Kempton, Illinois 60946 USA

www.adventuresunlimitedpress.com

Printed in the
United States of America

Cover by Terry Lamb

HIDDEN IN PLAIN SIGHT

Beyond the X-Files

TABLE OF CONTENTS

FOREWORD

The CIA, Fellow Travelers and Their Clandestine Underground & Undersea Lairs

I became aware of the CIA's grim nature early in life. I was born and reared just a short hop, skip and jump away from the CIA's main training facility at Camp Peary, in Tidewater Virginia. I've been by the place umpteen times over the years. It has always, in my childhood and adulthood, had a dark, glowering, foreboding sort of appearance from the outside, just driving past—the security fence, the brooding silence, the entryway. The place radiates a negative aura. How could it be otherwise?

In the late 1970s and 1980s, I read a suite of excellent books about the CIA, which I recommend to one and all:

In Search of Enemies: A CIA Story by John Stockwell
The Secret Team: The CIA and Its Allies in Control of the United States and the World by Fletcher Prouty
The Politics of Heroin: CIA Complicity in the Global Drug Trade by Alfred W. McCoy
The Crimes of Patriots: A True Tale of Dope, Dirty Money, and the CIA by Jonathan Kwitney

The first book, by John Stockwell, a former US Marine Corps officer and CIA agent, was very revelatory for me at the time, maybe 30 years ago when I first read it. Stockwell related how he became progressively more and more aware of the CIA's massive criminality, and of his own role in it, until he was led by his conscience to resign in disgust from the agency. The book got a fair amount of press at the time. As I recall the CIA tried to block it, and finally struck a deal where the CIA itself received the royalties from the book's sale!—or something like that. So much for 2nd Amendment, constitutional rights to freedom of the press.

The Secret Team, by Fletcher Prouty, first appeared in the 1970s. I avidly devoured it when I first read it in the 1980s. Prouty lays it all right out—the CIA by then had already metastasized into everything—Hollywood, the universities, Wall Street, Madison Avenue, publishing, radio and TV broadcasting, newspapers and magazines, politics, banking, the military, religion and much more. Its tentacles and operatives were everywhere, insinuating the CIA into everything. Prouty made the point that already as of the early 1970s, hundreds of compartmentalized units in the US military were *de facto* CIA operations, removed from effective control of the military chain of command. Think of the implications of that. Below, I will provide a concrete example of just one instance of what Prouty means.

The last two books, by Alfred McCoy and Jonathan Kwitney, have to do with the CIA's long-time, massive involvement in global narcotics trafficking operations, and associated massive money laundering activities. You could make the case that the CIA is the world's single largest narco-trafficking and dirty-money-laundering criminal cartel.

There's no need to put lipstick on a pig, and pretend that it is something that it is not. The CIA has been a homicidal, larcenous, lying, thieving, backstabbing, dirty-dealing, foul, vile organization virtually from its inception in 1947, when it was cobbled together from elements of the OSS, the United States' WW II intelligence agency, and the Gehlen Organization, the spy apparatus run by Adolf Hitler's wartime spy chief, General Reinhard Gehlen. Gehlen struck a deal with the Americans at war's end to deliver his Nazi spy organization to the American intelligence service, in exchange for clemency. Gehlen delivered the goods as promised, and several years later became the head of West Germany's intelligence service, the Bundesnachrichtendienst, or BND. The OSS and the Gehlen Organization were merged, yielding the modern-day CIA, meaning that from the very beginning the CIA has been a hybrid American-Nazi agency.

This is essential to understanding the world that we live in. World War II did not end as we have been taught; neither was it conducted as we have been taught. Beyond the hybrid Nazi nature of the CIA, for decades it has closely coordinated operations and intelligence with the Mossad, the Zionist intelligence agency, and other agencies such as the British MI-6, obviously the German BND, which Hitler's ex-spymaster Reinhard Gehlen went on to direct after WW II, and other agencies. If you find this improbable, bear in mind that after WW II well over one thousand Nazi technicians, engineers and scientists were secretly brought to the United States as part of Project Paperclip. They played an influential, crucial role in the manned space program of NASA and the ballistic missile programs of the US military. This is now publicly known by many people, though it was not well known in the 1940s and 1950s.

Another aspect of Project Paperclip that my research has uncovered has to do with the United States military's postwar

Underground Plant Program. In 1947 Project Paperclip requested Xaver Dorsch and three of his colleagues to be made available for the aforementioned "Underground Plant Program." Xaver Dorsch was at that time perhaps the premier expert in the world on underground base construction. At war's end he was the director of the Todt Organization, which was the Nazi military's civil engineering agency. It was the approximate equivalent of the US Army Corps of Engineers and the Navy Seabees, albeit that it was a nominally civilian agency. The Todt Organization was established by Fritz Todt, the father of the well-known German Autobahn system of highways. Fritz Todt was killed in an airplane crash mid-war, in 1942, and Xaver Dorsch was tapped to fill his position. Though Dorsch reported to Albert Speer, in 1944 Adolf Hitler asked Dorsch to begin a priority program of massive underground base construction, and to report directly to him on its progress, instead of to Speer. The war ended the next year and Xaver Dorsch was taken into United States military custody where he was extensively debriefed, and then requested by Project Paperclip in 1947. In 1951 Dorsch surfaced publicly again and founded the Dorsch Consult, now known as the Dorsch Gruppe, a large and well-known international engineering firm.

The point I am making is quite simple: the ongoing, underground bases program, the military and civilian aerospace programs and the CIA, all had important input and contributions from expert elements of the Third Reich in the years and decades after WW II's end.

The Fletcher Prouty Operation

As I have already mentioned, Fletcher Prouty was the

author of the eye-opening *Secret Team* book that provided so many troubling revelations about the CIA's secret dealings. I spoke with Prouty in the late 1990s and he revealed to me that beyond being a liaison officer during the height of the Cold War between the CIA and the US military, he had also been in charge of running the ratline that brought the Project Paperclip Nazis to the United States. That conversation was more than 20 years ago and I didn't know nearly as much, about a lot of things, as I do today. He confided to me that he and the others involved in Project Paperclip did not at first fully appreciate just how crazy the Nazi psychiatrists they brought over from Germany were. In retrospect, I believe he was alluding to the things the Nazis were doing for mind control techniques that were utilized in the then publicly unknown MK-Ultra and Project Monarch mind control operations of the CIA, which I and many others believe are ongoing in a major way to the present day.

Prouty also mentioned to me that the week that President Kennedy was murdered, he was abruptly ordered to the South Pole. Clearly he was ordered to the ends of the Earth to get him out of the way of the conspirators. Given his high level access to the CIA and military, the cabal involved in the plot obviously wanted him out of Washington, DC so that he could not interfere with the assassination.

At the time I spoke with Prouty I had already identified the US Army's Warrenton Training Stations in northern Virginia as the site of a CIA underground facility. Nominally the training stations were identified as, and run by the US Army as, US Army facilities. But in reality the US Army fig leaf was just a cover for what in actuality was a CIA underground facility. I asked Prouty about that facility. He didn't deny what I had discovered, but he replied that he couldn't tell me anything about the Warrenton Training Stations, or what was done there, because that got into the area of "Special Operations" and was classified, and he could not divulge classified information. Your guess is likely to be as good as mine as to what is really going on underground in Warrenton, Virginia.

It's not the only underground facility in Virginia. There are assuredly many others, a few of which are publicly known, such as Peters Mountain, in Albemarle County not far from Charlottesville, and Mount Weather, near Bluemont.

There is also a major underground facility beneath the Pentagon and very likely beneath the CIA headquarters in Langley, Virginia, in the Washington, DC suburbs, not to be confused with Langley Air Force Base at the other end of the state in Hampton, Virginia.

How Extensive Is the Underground System?

Very.

I have every reason to believe that what I know just scratches the proverbial surface.

Based on my research, and drawing on information that has been passed to me informally, I am fully confident that there are also underwater bases and tunnels, as well as underground bases and tunnels.

There are underground facilities and tunnels all over the United States. They can be anywhere—and are! Under forests, mountains, deserts, cities, etc. The technology to construct bases one mile deep is "comparative child's play" in the words of one expert who spoke to me off the record. I have been told that bases can even be as deep as several miles underground. I am not an insider, and I do not know all the technology that is available to the Deep State, but I have no doubt that what I have been told is true. The one-mile-deep figure is unquestionably state-of-the-art for the deep, underground base projects.

Many underground bases in the United States are known: the two in Virginia named above, Camp David in Maryland, Site "R" on the Maryland-Pennsylvania border, Cheyenne Mountain in Colorado, the NSA underground facility in Laurel, Maryland, the underground facility beneath the White House, the Manzano base in New Mexico, Area 51 in Nevada, the large facility beneath Offutt Air Force Base in Omaha, Nebraska, etc.

Without question there are many more, and many large facilities, that remain clandestine, Top Secret, compartmentalized and publicly unknown.

In terms of the most likely areas for these facilities in the continental United States the following have come to my attention:

a) the entire Appalachian region from New England all the way to northern Alabama and Georgia, including, but not limited to, the White Mountains in New England, the Interstate-81 corridor in Virginia, the eastern Tennessee region, West Virginia, and the mountains of Pennsylvania and western Maryland
b) the entire Rocky Mountain region
c) the Sierra Nevada in California
d) all of the desert Southwest, including Nevada and lapping

over into eastern and southern California

My research strongly points to a network of deep underground, high-speed, maglev (magnetic levitation) train tunnels connecting at least some of the underground bases. Based on everything I know, including hard and soft data, I consider it likely that a highly secret, high-tech, high-speed, underground transportation system has been built, likely in the mid-1970s to early 1990s time frame.

Based on my archival research and other information that has come to me, I also consider it likely that there are major tunnels and bases beneath the sea. Understand that once workers and construction equipment are down into the bedrock beneath the seafloor, the engineering challenges are not appreciably different from those encountered in a construction environment in solid rock thousands of feet beneath the surface of dry land. In both cases the operating environment is an enclosed environment in deep strata of solid rock.

I explore the technologies involved in constructing underground and underwater bases and tunnels in my three books:

Hidden in Plain Sight: Beyond the X-Files
Underground Bases and Tunnels
Underwater and Underground Bases

Suffice it to say that for more than half a century, the technology in marine engineering, civil engineering, mining engineering and petroleum engineering has permitted this type of construction to go forward. The techniques and machinery involved are very sophisticated and powerful.

As regards the underwater bases, the following locations are prime areas that are likely to have facilities:

a) the continental shelf off the eastern seaboard of the mainland United States, all the way from Florida to New England
b) the New England sea mounts in the North Atlantic off of northeastern North America
c) the Gulf of Mexico
d) the Caribbean Sea, especially (but not only) around Puerto Rico and Andros Island in the Bahamas
e) Lake Erie
f) the Channel Islands off of southern California
g) the Hawaiian Islands region
h) the Aleutian Islands region
i) the San Juan Islands, Vancouver Island region of western North America

What About the Funding?

Mind you that money is no issue, either to obtain the sums needed, or to hide the paper trail.

During my research I was privileged to have a lengthy conversation with a bona fide, high-level underground bases construction expert. I put to him the question as to how large, covert projects could be carried out with no paper trail of the funding.

To my astonishment, he proceeded to recite for me in detail

the funding for one of the clandestine projects on which he had been the lead project engineer. By the time he was finished I could scarcely comprehend what he had just told me. He laid out for me a convoluted, Rube Goldberg accounting scheme that involved offshore accounts, multiple United States government agencies, an Egyptian shell company, and much more that went right over my head. The money was moved around, in very large amounts, in what amounted to a high finance, international, inter-agency, money laundering shell game.

I was stunned. Rest assured, Dear Reader, that the banking laws and money laundering laws are written to oppress and control **you**—while the Deep State (the **literal** Deep State) does whatever it wants, and in a major, multi-million and multi-billion, even multi-trillion dollar way.

Yes, I mean that literally. Multi-trillions of dollars. They just skim it right out of the economy, defraud the public out of vast, unimaginable sums of money and do whatever they want: underground bases, underwater bases, secret, high-speed train tunnels, and God only knows what all else.

As Catherine Austin Fitts at the Solari Report – Solari.com has documented, the United States government cannot account for some $21 trillion in funding.

The Federal Government Can't Account for $21 Trillion. Does Anybody Care?

Where has the money gone? What has been done with it? Who really is running the United States government?

I submit to you that a lot of that money has quite probably been lavished on secret underground and underwater bases and tunnels.

As for who or what is really running the United States government? I don't know and neither do you, but clearly the Congress is not in charge, and neither is the President.

We are witnesses to a deceitfully staged, Punch and Judy political puppet show designed to distract our attention from the ENORMOUS craftily hidden theft that we are subject to, all while being told to: Vote! Pay taxes! Obey the law! Respect the government! Pledge allegiance to the flag! Support the military! Submit to official authority!

Is the CIA Involved?

You can be certain of CIA involvement. The CIA is massively involved in a wide range of Clandestine Operations, Covert Ops, Special Operations, call them what you like, and has been for the past 70 years.

Over the years I have heard the group that is running the domain of the secret underground and underwater bases referred to as the "Organization" or the "Company."

As it happens, for many years, the CIA has also been informally referred to as "The Company."

Why? Many explanations have been proffered, but certainly one of the biggest reasons has to do with the fact that the CIA is in business, extremely evil business to be blunt. Gun running/ weapons trafficking, human trafficking, narcotics trafficking, money laundering, and all of this on a massive, global scale.

It's all very violently, way out of control. And those are just the things we know that the CIA does, that can be and have been documented.

Given what my research has uncovered, and the CIA's

known, long-time involvement in the highly secure, underground base in Warrenton, Virginia, for me it is no stretch to believe that "The Company" I have heard referenced with respect to the massive, secret underground and undersea operations is, in fact, the CIA—or perhaps its Siemens subsidiary. Yes, that would be the same Siemens that played a role in the Third Reich and is still a major, international engineering firm, the same Siemens whose name keeps popping up in my underground and underwater bases and tunnels research. Not forgetting, of course, that the CIA was from its inception, a hybrid USA-Nazi agency.

I leave you with the thought that nothing is as we have been told. The political theater now underway in Washington, DC and elsewhere around the world is just a Punch and Judy puppet show to keep you distracted while the Deep State carries out its real agenda(s), its MASSIVE, $21 trillion, secret agenda(s).

I wish I could have another conversation with Fletcher Prouty. I know so much more as of August 2020 than I did when we spoke more than 20 years ago. But he's dead now and has taken his secrets with him to the grave.

But not to worry. We live on and it falls to us to uncover the whole, corrupt, unimaginably horrid, criminal mess and recover control of our lives, our society and our world.

Anyone with information for me can contact me c/o: Adventures Unlimited Press, the publisher of this edition of *Hidden in Plain Sight: Beyond the X-Files*.

—Richard Sauder, PhD
Quito, Ecuador, 16 August 2020

Chapter 1:

How Far Down

Nazi Engineers, Secret U.S. Military Bases, and Elevators To The Subterranean and Submarine Depths

In the preliminary stages of my research in the early 1990s, I had no informed idea of how deep below the surface underground bases could reach. By chance I had gone to hear a public talk by a man I did not know, on a topic that had nothing whatsoever to do with underground bases. However, during the talk he unexpectedly made an offhand comment that caused me to think that he knew something about secret underground facilities. A few days later, I telephoned him and asked if I could come by his office to speak with him. He consented to give me a little of his time, so I went by and asked him some questions, including about a specific location where I believed there was a secret underground base. He verified that there was a base there and told me that it was one mile deep.

At the time, that seemed improbably deep to me. I now understand, however, that it is well within the state of the art of the underground excavation and construction industries to build facilities one mile deep. In fact, facilities can be even deeper than that. Other information that has since been given to me has raised the question of facilities possibly as deep as 12 to 14 miles. Frankly, I do not know what the technological limits are in the super-secret, black budget realm of compartmentalized programs.

But I can make some fairly well educated deductions based on what I have been told and information gleaned from the open engineering and industrial literature. At a minimum, my investigation suggests that depths up to three miles are feasible, and that conclusion is based on a careful reading of the open scientific, mining and civil engineering literature. At one time I considered greater depths than that as unattainable, but no longer. While I don't know the limitations of the technology in the clandestine realm, based on everything I have heard over the years, it must surpass the limits found in the open technical literature. It could be that with classified improvements in materials science and mining engineering that bases can be sited many miles beneath the surface of the land or the seafloor. Frankly, I now surmise that we are dealing with a science-fiction-like reality that has been held back from public knowledge.

In that regard, it is clear to me that there are power structures on this planet that closely interface with, and yet remain separate from, the official power establishment of any nation as publicly presented in the mainstream news media. The sum total of all the many years of reading and research I have done, coupled with the myriad conversations I have had with an extremely eclectic selection of individuals, even raises the question as to whether some control structures and agencies for this planet may possibly extend off-planet. But even if that is not the case, we are certainly dealing at the least with trillions of dollars of off-the-books money, very advanced technology, an incredible infrastructure of secrecy, thousands and thousands of people who are in the know but say little or nothing about what they know, all of which revolves around projects of tremendous scope and complexity, carried out over a long period of time, and about which there is almost zero public knowledge.

It is unknown how many secret underground bases there are, but they surely do exist, there is no question of that. In 1987, Lloyd A. Duscha, the then-Deputy Director of Engineering and Construction for the U.S. Army Corps of Engineers, gave a talk at an engineering conference entitled "Underground Facilities for Defense – Experience and Lessons." In the first paragraph of his talk he stated that:

> After World War II, political and economic factors changed the underground construction picture and caused a renewed interest to "think underground." As a result of this interest, the Corps of Engineers became involved in the design and construction of some very complex and interesting military projects.[1]

A little further on he said:

> Although the conference program indicates the topic to be "Underground Facilities for Defense – Experience and Lessons," I must deviate a little because several of the most interesting facilities that have been designed and constructed by the Corps are classified.[2]

Subsequently Mr. Duscha went into a discussion of the Corps' involvement in the 1960s in the construction of the large and elaborate NORAD base buried deep beneath Cheyenne Mountain, in Colorado. And then he said:

> As stated earlier, there are other projects of similar scope, which I cannot identify, but which included multiple chambers up to 50 feet wide and 100

---

[1]Lloyd A. Duscha, "Underground Facilities for Defense – Experience and Lessons," in *Tunneling and Underground Transport: Future Developments in Technology, Economics and Policy*, ed. F.P. Davidson (New York: Elsevier Science Publishing Company, Inc., 1987), pp. 109-113.

[2] Ibid.

> feet high using the same excavation procedures mentioned for the NORAD facility.[3]

You will not find a franker public admission of the existence of secret, underground bases than that. It carries all the more weight coming from Lloyd Duscha, given his high position in the military-industrial complex.

The Project Paperclip, Nazi Connection

It is not accidental that Lloyd Duscha mentioned the U.S. Army Corps of Engineers' increased interest in underground bases and construction in the post-World War II period. In the closing stages of WW II, the U.S. military overran the Third Reich and confiscated a treasure trove of Nazi technology, engineering facilities and research laboratories. The American military also apprehended and detained large numbers of Nazi scientists, engineers and technicians, many of whom were brought to the United States under the auspices of the infamous "Project Paperclip" and integrated into American industry, research institutes and into the military and other official agencies.

The cases of such Project Paperclip notables as Wernher von Braun, and other "ex"-Nazis who were put to work building rockets and missiles for the military and NASA, are well known. The American political and military establishments were very impressed with Nazi V-2 and buzz bomb missile technology, and brought von Braun and his team to the USA at war's end to continue and advance the research and design they had done under Hitler's Third Reich. The development of the manned space program, satellite technology,

---

[3] Ibid.

deep space probes, intercontinental nuclear missiles and cruise missiles are all fruits of their work.

But Project Paperclip had another aspect which has received almost no publicity. And that aspect had to do with the underground facilities that the American military discovered when they entered the remnants of the Third Reich in the concluding stages of open military hostilities between the Allied and Axis powers in the European theater. I have two declassified Project Paperclip memoranda in my files that specifically request four men with expertise in underground construction, one of whom is Xaver Dorsch. The other men are not known to me, but Xaver Dorsch is. He was the director of the so-called Todt Organization in the closing phase of WW II. The Todt Organization was originally headed by Fritz Todt, whence the name. Fritz Todt was the designer and creator of the well-known German autobahn system of highways. The Todt organization served a somewhat analogous function in the Third Reich to that of the U.S. Army Corps of Engineers or the U.S. Navy Seabees in the present American military system. In other words, it was a civil engineering agency that supported the operations of the German military, though, unlike the case of the U.S. Army Corps of Engineers or the U.S. Navy Seabees in the American system, the Todt Organization was not a formal part of the German military structure. However, it did play a crucial role in the Nazi war effort and constructed many facilities for use by the German military and industry. Some of the underground structures that it built were impressively large and sophisticated.

In 1942 Fritz Todt was killed in an airplane crash and by war's end control of the agency had effectively passed to Xaver Dorsch. In the closing stages of the war, Hitler specifically tasked Dorsch with developing a series of huge, underground, industrial manufacturing facilities. Allied bombing raids were shredding German manufacturing and the Nazi command wanted to shift production underground, out of reach of American and British bombs. The collapse of the Third Reich ultimately put an end to those plans. I remain persuaded, however, that a great deal of what the Allies found when they went into Germany remains classified to the present day, and that includes the full extent of the underground facilities built by the Todt Organization.

Xaver Dorsch was taken prisoner by the American military on 7 May 1945. I have two moderately lengthy documents that he wrote for the Americans, evidently as part of his debriefing by them.[4] One was produced in 1946, the other in 1949-1950. The documents deal with the administrative structure and operational activities of the Todt Organization during the war. I strongly suspect that Dorsch supplied a great deal more information to his American captors during that period that has yet publicly to see the light of day. Xaver Dorsch may even have worked directly for the American military on the construction of secret, underground bases in the USA, just as he had done for the Nazis during the Third Reich. It is a fact that he was in American military captivity and was requested specifically by Project Paperclip.

[4] Xaver Dorsch, Diplomingenieur, Former Head of the Chief Office of the Organisation Todt, *The Organization Todt in France and Germany*, Steinlager Allendorf, 1 September 1946. Also, Xaver Dorsch, *Organization Todt (Dorsch Project)*, MS # P-037, English Copy, translated by G. Weber, edited by J.B. Robinson, reviewed by Capt. W.F. Ross, Foreign Military Studies Branch, Historical Division, Headquarters United States Army Europe, 1949-1950.

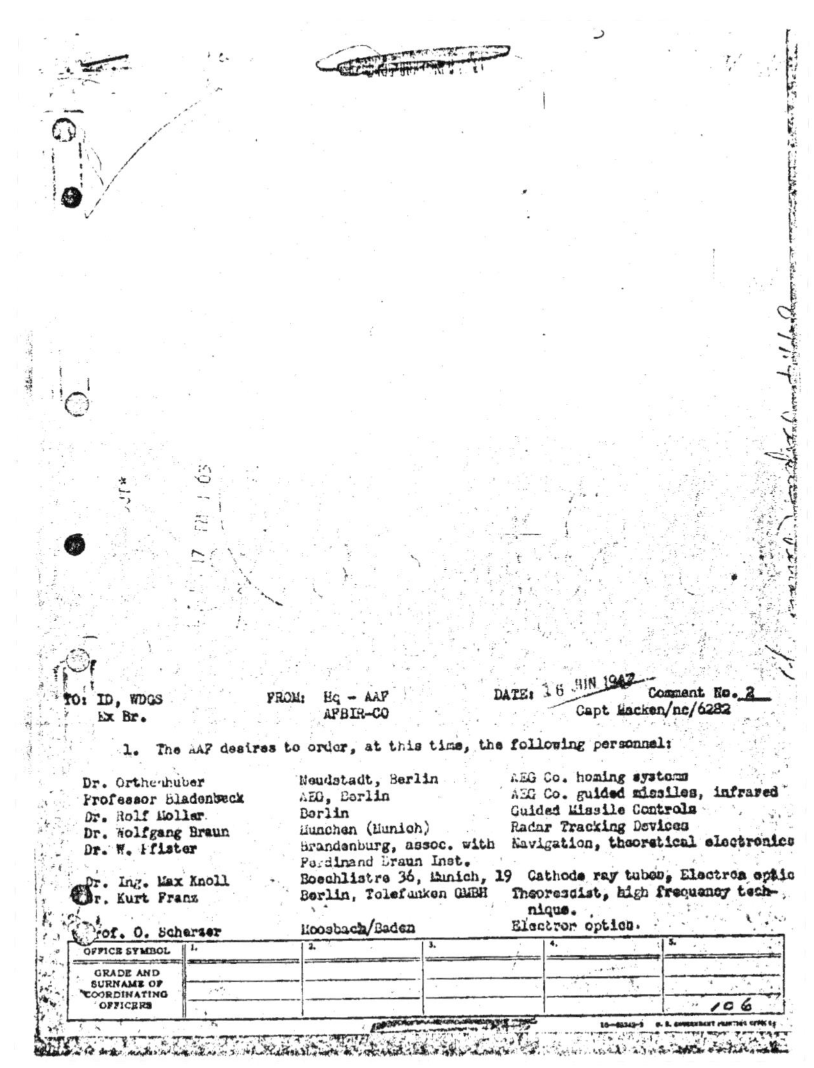

TO: ID, WDGS
Ex Br.

FROM: Hq - AAF
AFBIR-CO

DATE: 16 JUN 1947
Comment No. 2
Capt Macken/nc/6282

1. The AAF desires to order, at this time, the following personnel:

Dr. Orthemhuber	Neudstadt, Berlin	AEG Co. homing systems
Professor Bladenbeck	AEG, Berlin	AEG Co. guided missiles, infrared
Dr. Rolf Moller	Berlin	Guided Missile Controls
Dr. Wolfgang Braun	Munchen (Munich)	Radar Tracking Devices
Dr. W. Pfister	Brandenburg, assoc. with Ferdinand Braun Inst.	Navigation, theoretical electronics
Dr. Ing. Max Knoll	Boechlistre 36, Munich, 19	Cathode ray tubes, Electron optic
Dr. Kurt Franz	Berlin, Telefunken GMBH	Theorescist, high frequency technique.
Prof. O. Scherzer	Moosbach/Baden	Electron optics.

OFFICE SYMBOL	1.	2.	3.	4.	5.
GRADE AND SURNAME OF COORDINATING OFFICERS					106

Illustration 1-1a: Page one of a three-page declassified Project Paperclip memorandum, dated 16 June 1947. The memo was issued by Army Air Force Headquarters and explicitly asks for a list of German scientists and engineers, by name, in some cases along with their areas and fields of expertise and their addresses. *Source:* Project Paperclip file, NASA Historical Archives, Washington, D.C.

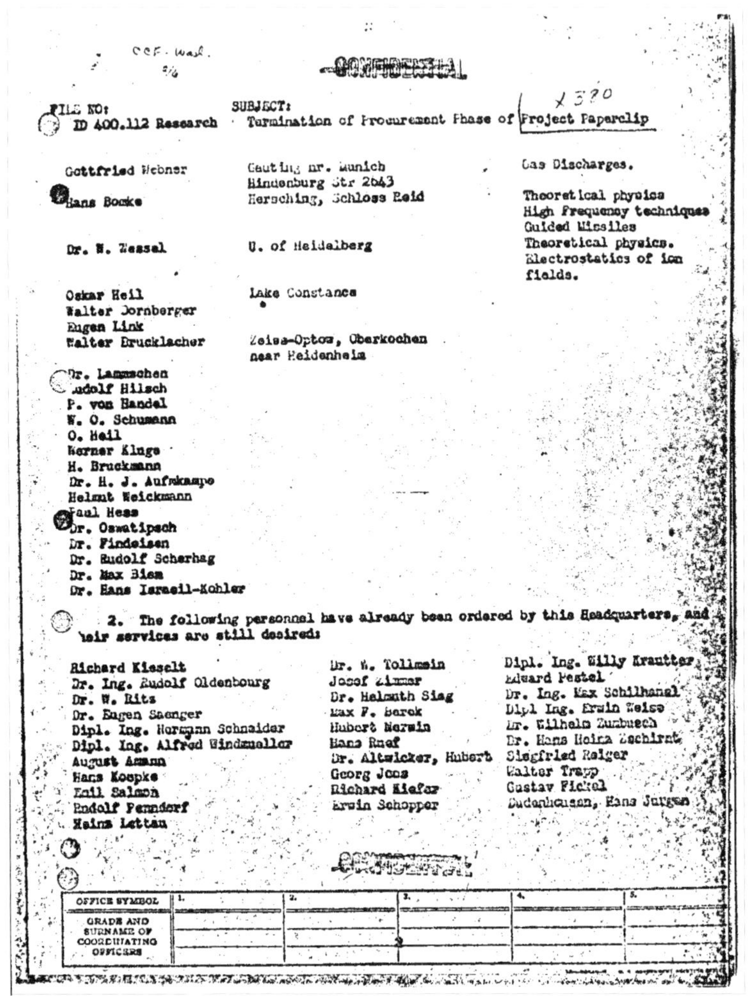

CONFIDENTIAL

FILE NO: ID 400.112 Research

SUBJECT: Termination of Procurement Phase of Project Paperclip

Gottfried Webner	Gauting nr. Munich Hindenburg Str 2643	Gas Discharges.
Hans Bocke	Herrsching, Schloss Reid	Theoretical physics High Frequency techniques Guided Missiles
Dr. W. Wessel	U. of Heidelberg	Theoretical physics. Electrostatics of ion fields.
Oskar Heil	Lake Constance	
Walter Dornberger		
Eugen Link		
Walter Brucklacher	Zeiss-Optos, Oberkochen near Heidenheim	

Dr. Lammachen
Rudolf Hilsch
P. von Handel
W. O. Schumann
O. Heil
Werner Klage
H. Bruckmann
Dr. H. J. Aufmkampe
Helmut Weickmann
Paul Hess
Dr. Oswatipsch
Dr. Findeisen
Dr. Rudolf Scherhag
Dr. Max Biem
Dr. Hans Israeil-Kohler

2. The following personnel have already been ordered by this Headquarters, and their services are still desired:

Richard Kieselt	Dr. W. Tollmein	Dipl. Ing. Willy Krautter
Dr. Ing. Rudolf Oldenbourg	Josef Zimmer	Edward Pestel
Dr. W. Rits	Dr. Helmuth Sieg	Dr. Ing. Max Schilhanel
Dr. Eugen Saenger	Max F. Barek	Dipl Ing. Erwin Weise
Dipl. Ing. Hermann Schnaider	Hubert Merwin	Dr. Wilhelm Zumbuech
Dipl. Ing. Alfred Windmueller	Hans Rnef	Dr. Hans Heinz Zschirnt
August Amann	Dr. Altwicker, Hubert	Siegfried Reiger
Hans Koepke	Georg Joos	Walter Trapp
Emil Salmon	Richard Kiefer	Gustav Fickel
Rudolf Penndorf	Erwin Schopper	Dudenhausen, Hans Jurgen
Heinz Lettau		

CONFIDENTIAL

OFFICE SYMBOL	1.	2.	3.	4.	5.
GRADE AND SURNAME OF COORDINATING OFFICERS					

Illustration 1-1b: Page two of a three page declassified Project Paperclip memorandum, dated 16 June 1947. One of the names listed here is Dr. Eugen Saenger, the father of so-called boost-glide rocketry, the forerunner of the space plane technology that we today know as the space shuttle. Note as well, the name of Walter Dornberger, the SS general who was in charge of the Nazi missile program at Peenemünde, and who was Wernher von Braun's commanding officer. Dornberger did come to the USA and worked for Bell Helicopter in New York. *Source:* Project Paperclip file, NASA Historical Archives, Washington, D.C..

FILE NO. SUBJECT:
ID 400.112 Research Termination of Procurement Phase of Project Paperclip

3. T-4, Air Materiel Command, has established a requirement for four (4) German technicians to act as consultants on the Underground Plant Program. The following four technicians presently working on this project for USAFE, A-2, have been recommended by that Command:

Xaver Dorsch	Helmut Ewald
Oscar Gabel	Walter Schrieber

The contract with an industrial firm to prepare a study on this project has not been approved by the Secretary of War and therefore, the action to bring the technicians to the United States is pending. The above list of scientists is subject to review by the contractor and may be changed to meet his requirements for consultants in specialized fields encountered in the underground construction study. Replacement selections will be made from the list of forty (40) engineers forwarded by USAFE to AC/AS-2.

4. Since it is virtually impossible under the research program to foresee all scientific personnel requirements as of any specific date; and since only those scientists are requested by the AAF whose services are essential to the national defense, it is hoped that this Headquarters may from time to time continue to request, and obtain, scientists whose services are urgently needed.

FOR THE COMMANDING GENERAL, AAF:

SIGNED

GEO. F. SCHULGEN
Brigadier General, U.S.A.
Chief, Air Intelligence Requirements Div.
Office of Ass't. Chief of Air Staff-2

OFFICE SYMBOL	1. AFBIR-CO	2. AFBIR	3. AFACB	4.	5.
GRADE AND SURNAME OF COORDINATING OFFICERS	Col Taylor / Capt Macken	Geo. F. Schulgen Brig. Gen., USA	Col. [illegible]		

Illustration 1-1c: Page three of a three page declassified Project Paperclip memorandum, dated 16 June 1947. On this page, under section three. the Air Materiel Command expresses a “requirement” for four German technicians to consult on its planned “Underground Plant Program”. Xaver Dorsch is the first name listed. *Source:* Project Paperclip file, NASA Historical Archives, Washington, D.C.

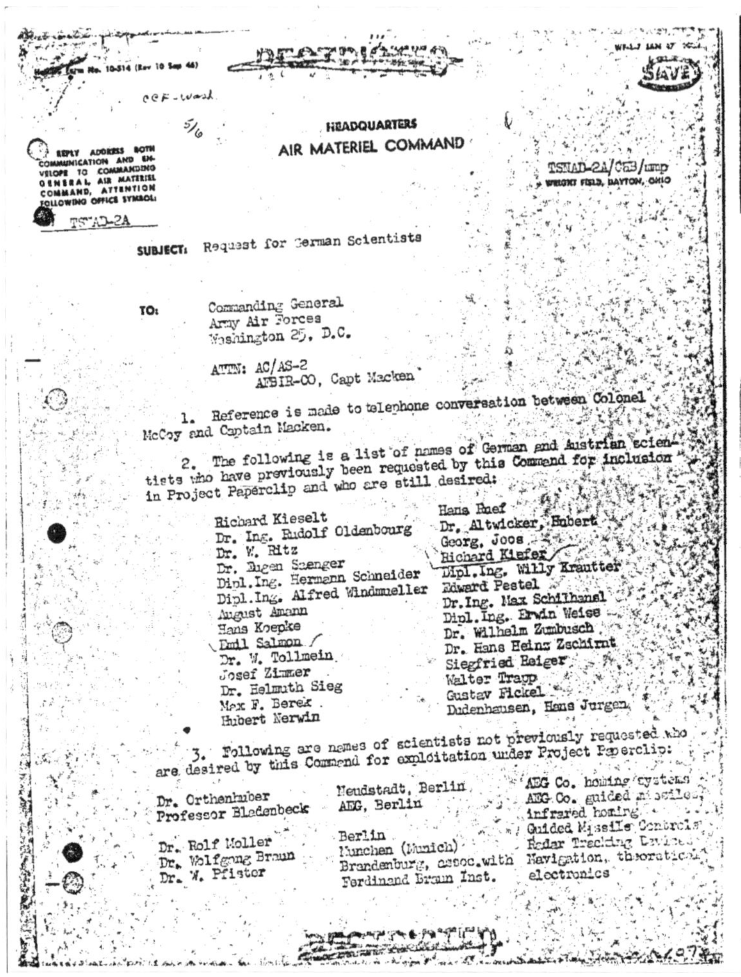
Form No. 10-314 (Rev 10 Sep 46)

SAVE

HEADQUARTERS
AIR MATERIEL COMMAND

REPLY ADDRESS BOTH COMMUNICATION AND ENVELOPE TO COMMANDING GENERAL, AIR MATERIEL COMMAND, ATTENTION FOLLOWING OFFICE SYMBOL:
TSNAD-2A

TSNAD-2A/CEB/ump
WRIGHT FIELD, DAYTON, OHIO

SUBJECT: Request for German Scientists

TO: Commanding General
Army Air Forces
Washington 25, D.C.

ATTN: AC/AS-2
AFBIR-CO, Capt Macken

1. Reference is made to telephone conversation between Colonel McCoy and Captain Macken.

2. The following is a list of names of German and Austrian scientists who have previously been requested by this Command for inclusion in Project Paperclip and who are still desired:

Richard Kieselt
Dr. Ing. Rudolf Oldenbourg
Dr. W. Ritz
Dr. Eugen Saenger
Dipl.Ing. Hermann Schneider
Dipl.Ing. Alfred Windmueller
August Amann
Hans Koepke
Emil Salmon
Dr. W. Tollmein
Josef Zimmer
Dr. Helmuth Sieg
Max F. Berek
Hubert Nerwin

Hans Ruef
Dr. Altwicker, Hubert
Georg. Joos
Richard Kiefer
Dipl.Ing. Willy Krautter
Edward Pestel
Dr.Ing. Max Schilhansl
Dipl.Ing. Erwin Weise
Dr. Wilhelm Zumbusch
Dr. Hans Heinz Zschirnt
Siegfried Reiger
Walter Trapp
Gustav Fickel
Dudenhausen, Hans Jurgen

3. Following are names of scientists not previously requested who are desired by this Command for exploitation under Project Paperclip:

Dr. Orthenhuber	Neudstadt, Berlin	AEG Co. homing systems
Professor Bladenbeck	AEG, Berlin	AEG Co. guided missiles, infrared homing
Dr. Rolf Moller	Berlin	Guided Missile Controls
Dr. Wolfgang Braun	Munchen (Munich)	Radar Tracking Devices
Dr. W. Pfister	Brandenburg, assoc. with Ferdinand Braun Inst.	Navigation, theoretical electronics

Illustration 1-2a: Page one of an undated, two page declassified Project Paperclip memorandum, circa 1947. This memorandum also requests Dr. Eugen Saenger, on whose early work the space shuttle concept is based. The first orbital shuttle flight was in 1981, and the first atmospheric testing was in 1977. *Source:* Project Paperclip file, NASA Historical Archives, Washington, D.C..

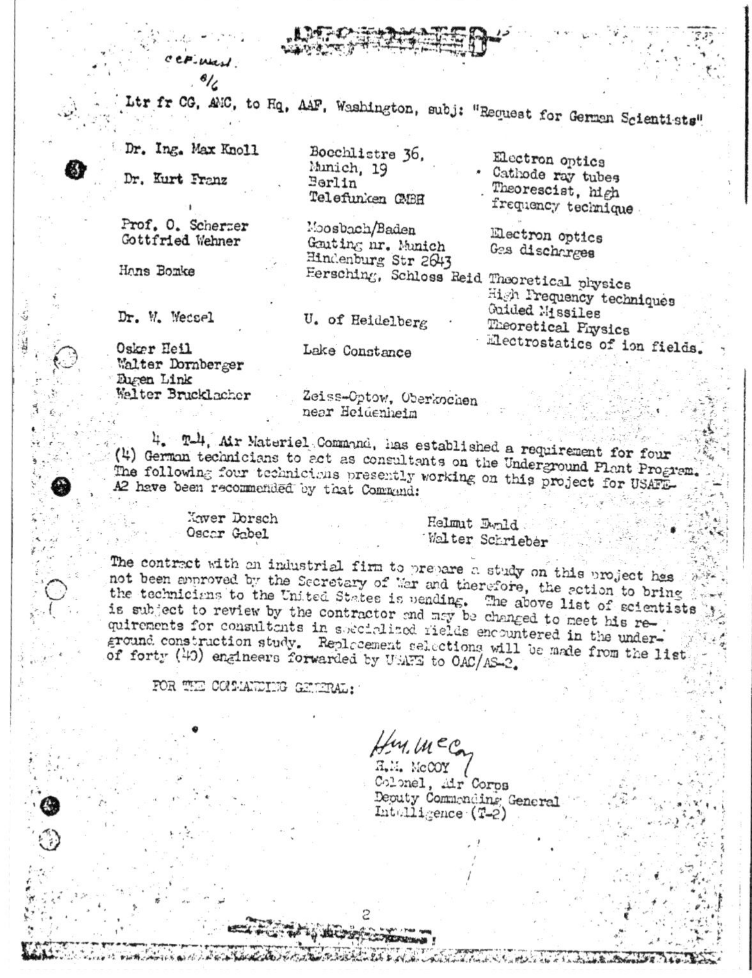

Ltr fr CG, AMC, to Hq, AAF, Washington, subj: "Request for German Scientists"

Dr. Ing. Max Knoll	Bocchlistre 36, Munich, 19	Electron optics
Dr. Kurt Franz	Berlin Telefunken GMBH	Cathode ray tubes Theorescist, high frequency technique
Prof. O. Scherzer	Moosbach/Baden	Electron optics
Gottfried Wehner	Gauting nr. Munich Hindenburg Str 2643	Gas discharges
Hans Bomke	Ferschinг, Schloss Reid	Theoretical physics High frequency techniques Guided Missiles
Dr. W. Wessel	U. of Heidelberg	Theoretical Physics
Oskar Heil	Lake Constance	Electrostatics of ion fields.
Walter Dornberger		
Eugen Link		
Walter Brucklacher	Zeiss-Optow, Oberkochen near Heidenheim	

4. T-4, Air Materiel Command, has established a requirement for four (4) German technicians to act as consultants on the Underground Plant Program. The following four technicians presently working on this project for USAFE-A2 have been recommended by that Command:

Xaver Dorsch
Oscar Gabel
Helmut Ewald
Walter Schrieber

The contract with an industrial firm to prepare a study on this project has not been approved by the Secretary of War and therefore, the action to bring the technicians to the United States is pending. The above list of scientists is subject to review by the contractor and may be changed to meet his requirements for consultants in specialized fields encountered in the underground construction study. Replacement selections will be made from the list of forty (40) engineers forwarded by USAFE to OAC/AS-2.

FOR THE COMMANDING GENERAL:

H. M. McCOY
Colonel, Air Corps
Deputy Commanding General
Intelligence (T-2)

2

Illustration 1-2b: Page two of an undated, two page declassified Project Paperclip memorandum, circa 1947. On this page, under section four the Air Materiel Command once again expresses a "requirement" for four German technicians to consult on its planned "Underground Plant Program". Xaver Dorsch is again the first name listed. Also of note here is the name of Walter Dornberger, the SS General who headed up the V-2 and V-1 programs for Hitler. Dornberger was brought to the United States and worked for many years for Bell Helicopter, in upstate New York. *Source:* Project Paperclip file, NASA Historical Archives, Washington, D.C..

During the course of my research I spoke with an expert who is personally familiar with some of the Nazi underground constructions in Europe and his observation to me was that they were very well made facilities. One of the best accounts of such facilities that I have found in the open literature is the anecdotal history related by Colonel Robert S. Allen that details what General Patton's army discovered when it entered Germany in the closing stage of World War II. Col. Allen revealed that Patton's forces found four large underground bases in the vicinity of the grim Nazi concentration camp near Ohrdruf, Germany; other underground facilities were reported in nearby towns. Col. Allen provided the following description:

> The underground installations were amazing. They were literally subterranean towns. There were four in and around Ohrdruf.... None were natural caverns or mines. All were man-made military installations. The horror camp had provided the labor. An interesting feature of the construction was the absence of any spoil. It had been carefully scattered in hills miles away.
>
> Over 50 feet underground, the installations consisted of two and three stories, several miles in length and extending like the spokes of a wheel. The entire hull structure was of massive, reinforced concrete. Purpose of the installations was to house the High Command after it was bombed out of Berlin. The Ohrdruf installations were to have been used by the Signal Communications Section. One, near the horror camp, was a huge telephone exchange equipped with the latest and finest apparatus. Signal Corps experts estimated their cost at $10,000,000.
>
> This place also had paneled and carpeted offices, scores of large work and store rooms, tiled bathrooms with both tubs and showers, flush toilets, electrically equipped kitchens, decorated dining rooms and mess halls, giant refrigerators, extensive sleeping quarters, recreation rooms, separate bars for officers and enlisted personnel, a moving-picture theater, and air-conditioning and sewage systems.
>
> Begun in 1944, the installations had been completed but never occupied.[5]

---

[5] Colonel Robert S. Allen, *Lucky Forward: The History of Patton's Third Army* (New York: The Vanguard Press, Inc., 1947).

The Regenwurmlager in Poland

Another spectacular example of Nazi underground engineering prowess was the subterranean Regenwurmlager complex that still sprawls for many miles deep beneath the countryside of western Poland. Several years ago Paul Stonehill wrote an eye-opening article about this site, replete with color photos, in *FATE* magazine.[6]

The Regenwurmlager was an obvious Nazi analog on the eastern German border region to the well-known Maginot Line that the French had built in their eastern border region, in that it also, like the Maginot Line, was comprised of many miles of underground tunnels and electric train lines connecting bunkers, fortifications and other critical military facilities.

Amazingly, the full extent of the huge complex is not known, even today. It is not known if either the Poles or the Russians have ever fully explored the many miles of tunnels and underground emplacements. According to Stonehill, the facility has many secret or hidden entrances, and an underground subway system with an electric train on rails that ran through a tunnel approximately 100 to 165 feet below the surface. On the surface, there were numerous military fortifications and bunkers crammed with weaponry, connected by secret passages to a sprawling labyrinth of tunnels below ground, said to be 30 miles or more in length. Stonehill reports that Adolf Hitler is alleged to have visited the Regenwurmlager in 1937, riding in on the underground subway train.[7]

---

[6] Paul Stonehill, "Secrets of the Regenwurmlager," *FATE* vol. 55 no. 10 issue 631 (November 2002): 28-33.

[7] Ibid.

It happens that the Regenwurmlager complex of military bunkers and underground tunnels and subway trains was located in territory that was overrun by Soviet troops in the closing stages of World War II, so the American military may never have examined the facility. However, I consider it very likely that Xaver Dorsch would have had personal knowledge of this facility, and equally likely that he briefed his American captors on it.

This book cannot provide an exhaustive treatment of Nazi underground facilities, but the examples provided suffice to demonstrate that 65 to 75 years ago, Nazi engineers already had a sophisticated capability to construct elaborate, large facilities underground. The American military understood this clearly at war's end, and urgently wanted to bring Nazi experts to the USA to build underground facilities here also.

By the time you finish this book, you should understand that this mission was accomplished. While we still do not know the full scope of what the Nazis did underground decades ago, we also still do not know the full extent of what the American military, and other agencies and corporations, have built underground right here in the USA, and elsewhere in the world. What can be said with certainty is that an extensive program of secret underground construction began in earnest in the years following WW II and that program continues in effect to this day.

Where Are the Bases?

In the continental USA there are numerous underground bases. I will briefly list just some of the bases that my research has uncovered over the last 17 years. There are assuredly many more than those listed here.

Maryland

- *Camp David*, the Presidential retreat, is immediately adjacent to Catoctin Mountain Park. On the surface there are some cabins and lodges, and conference rooms for Presidential use. But most of this facility runs like a labyrinthine anthill under Catoctin Mountain, just west of Thurmont, Maryland, along Rte. 15 between Frederick, Maryland and Gettysburg, Pennsylvania. The facility is operated and maintained by the U.S. Navy Seabees. The U.S. Marines provide military security. I have spoken to an individual who used to work there on a classified maintenance contract. My source informed me that the underground portions of Camp David are so extensive and elaborate, and there are so many miles of secret tunnels, that it is doubtful that any one person would have a complete map of the facility in his or her mind. In other words, we hear or read the words, Camp David, in the mainstream news media, but understand virtually nothing of what really happens there, or what the full nature of the installation really is.

- *Fort Meade*, near Laurel, Maryland, midway between Washington, D.C. and Baltimore, Maryland. My research, from a variety of sources, indicates that there is a huge underground maze beneath Fort Meade. Fort Meade is operated by the U.S. Army, but the National Security Agency (NSA) has its largest, publicly known operations center there. The NSA has literally acres upon acres of super computers underground at Fort Meade, stacked level upon level, going down and down and down, like a gargantuan subterranean stack of hi-tech digital pancakes. The NSA is notorious for the ECHELON spying program and other unconstitutional spying activities. It is something like a high-tech KGB or Stasi surveillance agency, spying on the American population's electronic and digital communications.

■ *FEMA Alternate Underground Command Center*, off of Riggs Road, not far from Olney, Maryland, to the north of Washington, D.C. When I visited the place back in the 1990s, only a few shabby, almost dilapidated looking buildings were visible through the security fence. A few vent pipes poked above the ground here and there. There were a few largish antennae and radio masts visible. The casual onlooker would probably just drive right by, unaware that the real activity was taking place below ground. I spoke with one man who had been in the facility and he described a multi-level base crammed full of sophisticated electronic gear. He had been to level seven underground, but did not know if the base extended deeper than that, or whether it was connected via deep tunnels with other underground bases in the region.

■ *Site R*, also known as *Raven Rock* or the *Underground Pentagon*, is about 6 miles north-northeast of the Camp David facility, not far from the Maryland-Pennsylvania state line. This enormous facility is the alternate underground command center for the Pentagon. In recent years Dick Cheney reportedly spent a lot of time underground at Site R. I have heard that Site R and Camp David are connected by underground tunnels, and I am inclined to believe the rumors are true. Site R is now under the command of Fort Detrick, in Frederick, Maryland, to the northwest of Washington, D.C.. Fort Detrick is the U.S. Army's infamous biological warfare research facility.

Virginia

■ Mount Weather, near Bluemont, Virginia on Rte. 601, is the major underground command center for the Federal Emergency

Management Agency. FEMA has a 400 acre complex on the surface of the mountain. The underground base is very large, and actually qualifies as a high-tech, subterranean town. This base dates to the 1950s.

■ *The Pentagon*, in northern Virginia, the national military command center, across the Potomac River from Washington, D.C.. The Pentagon was built directly on the site of Robert E. Lee's former plantation. I have been told that there are multiple underground levels beneath the Pentagon. Just how deep the complex goes is unclear, but elsewhere in this book I discuss plans (dating to the 1960s) to build a very deep base at the 3,500 foot level. I consider it likely that something like this has been done.

■ *U.S. Army Warrenton Training Stations A and B*, in the near vicinity of Warrenton, Virginia are ostensibly U.S. Army facilities. But in reality, my research showed that there is a CIA presence there. As for what sorts of operations take place there – who knows? I visited these places in June of 1992 and found Station A on Rte. 802; Station B is on Bear Wallow Road. There are also Stations C and D elsewhere in the region, which I did not visit. In the course of my research, I called up Col. Fletcher Prouty, the well-known author of the book, *The Secret Team.* For many years, at the height of the Cold War, Prouty was a liaison officer for the Air Force, helping the CIA with its clandestine activities worldwide. I surmised that if anyone ought to know something about what was going on at Warrenton it should be Fletcher Prouty. So I put the question to him point blank, and asked him what the CIA was doing underground at a U.S. Army training installation in Warrenton, Virginia. To begin with, he confirmed that the CIA was indeed present in Warrenton, and was using the U.S. Army as a cover. I asked him what was going on underground. He responded, "Well, you have to understand that that gets into the realm of Special Operations and that's classified." And that was all that I could pry out of him. I suspect the entire locale is tunneled out underground. The palpable Alice in Wonderland aura exuded by the Warrenton U.S. "Army" Training Stations continues to linger in my memory, even with the passage of 17 years.

■ Before leaving Virginia, I also want to say something about another low profile CIA facility. *Camp Peary* is located just a few miles from Colonial Williamsburg. It is sometimes referred to as "The Farm" in popular parlance. Though it is hard to find out much about the base, enough is known to say with confidence that Camp Peary is the CIA's main training and operations base in the USA.

As with the Warrenton Training Stations, the CIA uses a thin U.S. military cover at Camp Peary; in this case, maps indicate that Camp Peary is a U.S. Navy Reservation. I am a native Virginian from the Tidewater region, and spent my early childhood in a community just 20 miles away from Camp Peary. In the 1970s I attended the College Of William & Mary, in Colonial Williamsburg. Camp Peary lies on the York River, just to the east of the town of Williamsburg, on the northern side of Interstate Highway 64. I have driven past the place numerous times over the years and it has always seemed darkly brooding to me. Immediately to the southeast of it lies the U.S. Navy's Cheatham Annex, a major weapons supply depot for the U.S. Navy's Atlantic Fleet, part of the Yorktown Naval Weapons Station.

This interests me, because many years ago I spoke with a woman whose father worked at the Cheatham Annex/Yorktown Naval Weapons Station complex in the years after World War II. He was a construction worker, and at that time, about 60 years ago, the U.S.

Navy was building a facility deep below the water line. The whole site lies only a few feet above sea level, so the water table is very close to the surface. The Navy used high powered water pumps to instantly pump out the brackish ground water that was rushing in, to keep the excavation from filling with water. My acquaintance told me that her father said they used a fast setting concrete. She did not know the purpose of the facility. The instructive part of the story is that even then, 60 years ago, military engineers had the capability to construct facilities below the water line, using high powered pumps and quick setting concrete.

This Navy facility, constructed 60 years ago, lies immediately adjacent to the CIA's secretive Camp Peary base (which is nominally administered by the U.S. Navy). Recalling that the Camp David Presidential retreat in Maryland is also administered by the U.S. Navy and lies above a major underground base, I conclude that there is a high probability that the Camp Peary/Cheatham Annex area along the York River is underlain by an underground complex, too.

Washington, D.C.

■ *The White House* has a very large, deeply buried facility underneath it. One of my close friends was taken down into this facility during the Lyndon B. Johnson administration in the 1960s. She entered an elevator in the White House and was escorted straight down. She believes that the elevator went down 17 levels. When the door opened underground she was escorted down a corridor that appeared to disappear to the vanishing point in the distance. Other doors and corridors opened off of that corridor. That was what Washington, D.C. was *really* like underground almost half a century ago.

■ My research indicates that Washington, D.C. has a veritable labyrinth of tunnels beneath it. Some of the tunnels are publicly known, such as the Metro tunnels and the underground train tunnels that members of Congress use to travel from their office buildings to the Capitol building. As I discuss elsewhere in this book, I have been told and read stories of other tunnels that are more secret. In light of the available evidence, I incline to the view that these stories contain an appreciable degree of truth.

Texas

■ *Medina Annex*, at Lackland Air Force Base, in San Antonio. This facility lies in southwest San Antonio, on the south side of Rte. 90, immediately west of the junction of Interstate 410 and Rte. 90. This is one of the original Q Areas built by the Pentagon back in the late 1940s and 1950s for the storage and assembly of nuclear weapons. Local lore has it that the underground portion is very large. I have been told that the underground base is very deep and cold, for whatever reason. The National Security Agency (NSA) also has a major presence at the Medina Annex with thousands of personnel. I have on occasion driven by the Medina Annex on Rte. 90 and the base has a very dark, ominous presence.

■ *Camp Bullis*, in northwest San Antonio, Texas, not too far from Interstate Highway 10. This is an Army training base, immediately adjacent to neighboring Camp Stewart, another Army base that keeps a much lower profile. I have been told that the Camp Bullis/Camp Stewart reservation is the site of an underground base.

■ *Fort Hood*, near Killeen, Texas, 70 miles north of Austin. This was also the site of one of the original Q Areas. I have been told that the area of Fort Hood, and of the former Killeen and Gray Air Force Bases, is the site of a secret underground complex. I communicated with one ex-Fort Hood soldier whose duty area was at a checkpoint two miles inside a tunnel leading to an underground area, near the former Gray Air Force Base.

Nebraska

■ *Offutt Air Force Base* has had a major underground facility for decades. During the Cold War it was the underground command center for the Strategic Air Command. George W. Bush flew there for protection on Air Force One during the 9/11 attacks.

New Mexico

■ Since the late 1940s there has been a major underground base at *Kirtland Air Force Base*. The base is beneath a foothill of the Manzano Mountains. It was originally constructed as a super-secure nuclear weapons assembly and storage facility for the military. Today the base is used by other agencies. The Air Force has built another, state of the art facility for nuclear weapons storage at another location at Kirtland AFB. The Department of Energy's Sandia National Laboratory is immediately adjacent to Kirtland Air Force Base, on the southeast border of Albuquerque.

I talked with one of the men who helped to build the Manzano base in the years after World War II. He spent his entire career as a hard rock miner for federal agencies, working on underground projects all over the western USA. While excavating the Manzano base, security was extreme. The miners were blindfolded when they were transported to and from their work site. When they were taken from the area where they were working to another part of the facility they were always blindfolded. The practical result of this procedure was that not even the miners who built the underground base knew its layout. All they ever saw was the immediate chamber or tunnel section they were currently working on. It was a strictly compartmentalized project in every sense of the word. I suspect that this facility has been expanded over the years.

- *Los Alamos* is one of the U.S. Department of Energy's national research laboratories, to the west of the state capitol of Santa Fe. I have been informed that the underground work space beneath Los Alamos is even greater in extent than the sprawling surface facility, reaching as much as one mile deep in its farthest reaches. Two of the main missions of Los Alamos are nuclear and genetic research. Presumably classified projects related to these fields are among the secret activities carried out underground.

- *White Sands Missile Range*, in south-central New Mexico, is the largest military base in the USA. It was the site of the first test explosion of an atomic bomb in 1945. White Sands was the launch site for test firing of captured Nazi V-2 rockets after the military defeat of the Third Reich. Today it is still an operational military missile range, and also serves as an alternate space shuttle landing site for NASA. Underground facilities beneath White Sands are reportedly devoted to cutting edge research in lasers and conscious supercomputing. The sum of my research strongly indicates a *major* underground component at White Sands. Nothing would surprise me where White Sands is concerned. Nothing whatsoever.

Nevada

■ The notorious *Area 51* of UFO fame and the *Tonopah Test Range/Nevada Test Site/S-4/Nellis Air Force Base* reservations in southern Nevada are all in relatively close proximity to each other. My research and sources point to major, massive underground facilities in this region. A great deal of ultra-sensitive military R&D takes place here. Much popular lore has arisen around the theme of captured UFOs held in secret by the American military at Area 51. In my estimation, these stories are probably true. Ryan Wood and other researchers have accumulated a great deal of evidence over the years that indicates that American military agencies are lying and that they do have captured UFOs. Area 51 is reportedly one of the places where this sort of technology is sequestered and studied. It is my informed opinion that White Sands Missile Range in New Mexico is another site for the study and R&D of exotic technologies.

Colorado

■ *Cheyenne Mountain*, near Colorado Springs, was the major underground command center for NORAD during the Cold War. It is a very deep, highly secure base. While Cheyenne Mountain is still in use, most of NORAD's daily operations have now been switched to Peterson Air Force Base in Colorado. This underground base has been featured in many motion pictures and television programs.

■ *The Denver Federal Center*, on the western edge of the Denver metropolitan area is the location of a FEMA underground command center. Many people are concerned about the Denver airport as being the site of an alleged underground base, but for my money the Denver Federal Center installation is more important.

Tennessee

■ *Naval Support Activity Mid-South* is a huge U.S. Navy base, in Millington, Tennessee just 21 miles north of Memphis. It is an enormous facility, covering thousands of acres. I have been told that there is a deep underground facility beneath this base. Remember that the U.S. Navy operates and maintains Camp David, which is also the site of a major underground facility.

California

■ *China Lake* is a massive U.S. Navy R&D reservation in eastern California that has long been rumored to be the site of a major underground base. While I can offer no direct proof, short of taking a live video crew underground to take a look around, the bulk of my research comes down on the side of a massive complex beneath the China Lake Naval Weapons Center, along Rte. 395, in the general vicinity of the town of Ridgecrest.

■ Reported massive, very deep underground installations are said to run out of Nevada and into eastern California. So much of the region is controlled by Federal departments and agencies, whether the Pentagon, Department of Interior, Department of Agriculture, Department of Energy, or unknown agencies, as in the case of Area 51, that that entire area of the country is essentially one large Federal Government Multi-Agency Reservation. Numerous military bases sprawl across the landscape, and then there are National Forests, National Monuments, and other lands under Federal control. This provides the Feds with the opportunity to move almost any necessary personnel and equipment around without attracting undue attention.

* * *

This list should give you a brief idea of what is under our feet. I am *sure* that there are many more facilities than those I have just discussed above. Please take due notice that many of these installations are gargantuan. An expert source once described to me an excavated underground space he was familiar with that was inside a mountain: it was approximately 1,000 feet high, 600 feet wide and 1,000 feet long. He wanted to give me an idea of the state of the art in hard rock, underground excavation. I was impressed, as I suspect you would be too.

We are faced with a global system that is so secretive and so wildly out of control that the vast majority of us have no earthly clue as to what is going on, on this, the planet that we inhabit.

The Elevator to the Sub-basement

During the course of my research I have spoken to several people who allege to have been escorted down into secret or highly secure underground facilities. A long, deep elevator ride is a common feature of their accounts.

Right about here, I imagine you are thinking to yourself: "How deep do the elevators go?"

The short answer is that the documentation in the open literature suggests that the answer would be anywhere from hundreds to thousands of feet, based on the capabilities of high-rise elevators built by companies such as Otis and ThyssenKrupp. The testimony of the people I have spoken with comports well with that information. I have also run across softer, undocumented information from time to time that suggests that some facilities go even deeper, several or even many miles deep.

I have little interest in doing an exhaustive survey of the elevator literature, but let me provide you just a few brief ideas of the state of the art in that industry. Keep in mind that an elevator system that is installed in an interior elevator well and goes to the top of a high building could just as easily be installed in a vertical shaft that goes straight down underground. Multiple elevators could be staggered on multiple levels to go down and down, as deeply as you desire. The same technology can be used in either case, whether you are going up, or *down*.

Otis Elevator Co. announced plans in 2002 to install high speed Skyway elevators in the new 880 foot tall Eureka Tower in Melbourne, Australia, that rise at 1,800 feet per minute. Otis also planned to install its innovative Gen2 flat belted elevator hoist system on the same project, that utilizes a permanent magnet machine, which is quieter and takes up less space than the traditional elevator machine room.[8] Another article I encountered referred to a "mile-high, multidirectional elevator being developed for Otis Elevator Co. for use in extremely tall buildings of the future."[9] Obviously, if a company can develop a mile-high elevator, it can also develop a mile-deep elevator. Toshiba /G.F.C. Elevator installed two high-speed elevator cars in the Taipei 101 tower that zoom up to the 89th floor observation deck at an ear-popping 3314 feet per minute.[10]

---

[8] "From Zero To 300 In 60 Seconds Otis High-Speed Elevator Systems Selected For Landmark 88-story Melbourne Tower," http://www.otis.com/news/newsdetail/0,1368,CLI23_NID11699_RES1,00.html, 2002.

[9] "Micro Craft Does Major Business," http://www.microcraft.com/Inside Outlook/WN_tenn.htm, 2002.

[10] "World's Fastest Elevator," http://www.popularmechanics.com/science/extreme_machines/2004/3/elevator/print.phtml, 2004.

That will give you a little idea from the open literature of how fast and how high modern elevators can go – very high and very fast!

As I dug into the elevator literature a little more I happened to read a book by Jason Goodwin about the Otis Elevator Company. The book is entitled, *Otis: Giving Rise to the Modern City*. I did not know who Mr. Goodwin was until I read his book, but I quickly came to understand that he is an extremely knowledgeable man in the field, having worked for Otis in a variety of important positions for 37 years, and then having formed his own elevator consulting company after his retirement from Otis. I read the book carefully and I found out a few interesting things for my research:[11]

1. Elevators can be located in, and I quote verbatim: "...the legs of a deep-sea oil production or oil production platforms, and in many other extraordinary locations."
2. There are very large platform "lifts for extremely special applications" that may use a variety of lifting technologies, including "screw jacks."
3. And in a very brief summary at the conclusion of his most informative book, Mr. Goodwin briefly mentions the elevators "that are never mentioned but are needed to service the extensive infrastructures that make the cities run – the power plants, refineries, factories, and *underground facilities* (my emphasis).

Amen brother, *underground facilities* and the elevators that service them. That is what this book is about. Mr. Goodwin doesn't really elaborate about that topic to any appreciable extent. But to his credit he does indeed mention elevators that service *underground facilities* beneath the nation's cities. His mention of the "extremely special

[11] Jason Goodwin, *Otis: Giving Rise to the Modern City* (Chicago: Ivan R. Dee, 2001).

applications" that require large platform lifts instantly caused me to reflect on David Adair's description (elsewhere in this book) of a mammoth, football-field-sized platform elevator supported by huge screws the size of giant sequoia tree trunks that took him underground at Area-51 in Nevada. As for the mention of elevators in the legs of giant oil production platforms, and in "many other extraordinary locations," the implications for access to manned undersea bases is clear. The Gulf of Mexico and North Sea are dotted with myriad oil production platforms, with their legs sunk into the seabed. Any of those platforms could potentially serve as an entry point to the sub-seafloor environment via elevators in their legs, which would permit personnel to travel to the platform and then travel down below the seafloor. I strongly suspect this is the case in some instances. In his book, *Alien Contact*, Timothy Good mentions a NORAD offshore platform in the Gulf of Mexico that superficially resembles an oil production platform, but is *not* in reality an oil platform. The mission of the strange platform seems to be to serve as a sort of monitoring station to watch the air space of the USA for UFO activity.[12] Of course, this begs the question as to how many ostensible offshore petroleum production platforms are not really oil rigs at all, but only resemble oil platforms as they perform other functions entirely? How many such cases are there? Only the oil companies and major governments would know the answer to that. The average person absolutely lacks the means to ascertain those facts. But thanks to Jason Goodwin's book, we know that oil platforms have elevators in their legs. Please note that Mr. Goodwin is not alleging that elevators in oil platform legs can go below the seafloor and access undersea, manned installations. That is a

---

[12] Timothy Good, *Alien Contact: Top-Secret UFO Files Revealed* (New York: Quill, William Morrow, 1993).

conclusion that I am drawing from the available evidence. I am not saying that every offshore oil production platform has elevators in its legs that access the sub-seafloor environment. However, it is clear that some platforms certainly could serve in this way as entry points to manned, undersea installations.

Chapter 2:

Letters From the Underground Mail Bag

Early on in my research on underground bases and tunnels, I began to receive occasional postal letters, e-mail and first-hand personal anecdotes from my readers and public lecture audience members. As the years have gone by, the letters, e-mails and personal anecdotes continue to trickle in. As you might surmise, some of the mail and stories have been most interesting. They have been so interesting, in fact, that I want to present some of that mail and information for you in this chapter. I am sure that it will interest you just as much as it has me.

I believe there is a substantial kernel of truth in the accounts you are about to read. I also want to add that, although I absolutely lack the time and money to travel around the country and world checking out the particulars of every last sensational story that crosses my path, I have, nevertheless, strongly relied on my research experience and knowledge of the subject matter to sift out obvious tall tales and wild stories. What remains for you to read is the credible remnant that I judge to have the ring and feel of true fact.

I have often been struck by the carefully staged and maintained illusions that we accept as the consensual reality of everyday life. The propagandistic news on television and radio, and in the newspapers and news magazines, the many false pronouncements by government, deeply entrenched journalistic and scientific censorship, and other sources of falsehood and deceit flowing from educational and religious institutions serve to foster the mass delusion that everything *is*, in reality, as it *appears* to be.

Nothing could be further from the truth.

Especially in the nether realms of underground and underwater bases and tunnels, where the operative rule is: *Out of sight, out of mind!*

So without further ado, on to the underground mail bag!

Subterranean Labyrinth Beneath the NSA at Fort Meade, Maryland

Deep beneath the National Security Agency headquarters at Fort Meade, Maryland, midway between Washington, D.C. and Baltimore, Maryland lies a Top Secret subterranean complex. In his masterful exposé of the NSA, *The Puzzle Palace*, James Bamford alludes to the mammoth computing center that lies beneath the NSA:

> ...the NSA's enormous basement, which stretches for city blocks beneath the Headquarters-Operations Building, undoubtedly holds the largest and most advanced computer operation in the world.[1]

This is the digital machinery the NSA uses to eavesdrop on electronic communications the world over, including in the United States. James Bamford says that this "basement" full of super-computers

---

[1] James Bamford, *The Puzzle Palace: A Report on America's Most Secret Agency* (Boston: Houghton Mifflin Company, 1982).

extends for "city blocks." I am sure you will agree that that is impressively huge.

One of my sources who had been briefly stationed at Ft. Meade during the 1970s accidentally stumbled into this mammoth "basement" and fully corroborates Bamford's account. Here is his account:

> I...was transferred to the NSA on a special assignment for a few weeks. One day, while working in the "building," I noticed a door next to the stairs by the east entrance where I came in at. There were no markings on it so I felt I could enter and not get in trouble.
>
> I opened the door and it was a stairwell leading down. I went over to the edge and looked down between the railings. I didn't count the number of floors down, but I had the feeling it was about 15-20 floors. It was a long way to the bottom.
>
> I walked down one flight and there was a door on the opposite side which would be under the ground level on the east side of the building. I opened the door and stuck my head thru (sic) and looked left and right and saw a tunnel which ran clear out of sight in both directions. It was definitely much further than the area covered by the building and parking lot at ground level. There were doors along the opposite walls spaced about 30-40 feet apart. The tunnel had concrete walls and floor with colored markings.
>
> I looked around to see if there were any cameras and saw none, but I was not entirely comfortable being there. But I decided to check out a couple of more floors so I walked down another level and there was another door. I opened it and looked in and saw the same layout – doors and the tunnel going as far as I could see in both directions. I went down one more floor and looked in and saw the same as the first 2 floors. In this one I saw a golf cart with 2 people on it in the distance heading my way so I left.
>
> Even these 3 tunnels were extensive. If the other levels below had tunnels also there is a massive tunnel complex under Ft. Meade.

Now remember that this account is from the late 1970s. It would not surprise me if the complex beneath the NSA has grown even larger, deeper and more elaborate in the intervening years. Indeed, I surmise that many important government agencies and military bases are likely to have comparable, huge, underground facilities beneath them.

The Underground Base at China Lake, California

For years the United States Navy has had a huge base at China Lake, California. In my first book, *Underground Bases and Tunnels*,[2] I presented military documentation from 1964 indicating that the United States Army Corps of Engineers was considering constructing a huge underground cavity 4,000 feet deep beneath the sprawling China Lake facility.

As it happens, after giving a public talk a couple of years ago, I was approached by a man who told me he had been a uniformed member of the United States Navy. We chatted for a while and when he mentioned that he had spent some time at China Lake my ears perked up. I asked him if there was an underground facility at China Lake. He said that indeed there is, and that it is impressively large and deep. I asked him if he had ever been in it, and he said that he had, though not to the deepest levels. I asked him how deep the deepest part extended.

He looked at me soberly and said very quietly, "It goes one mile deep."

I then asked him what the underground base contains. He replied, "Weapons."

I responded, "What sort of weaponry?"

And he answered without pausing, "Weapons more powerful than nuclear weapons."

---

[2] Richard Sauder, *Underground Bases and Tunnels: What is the Government Trying to Hide?* (Kempton, Illinois: Adventures Unlimited Press, 1995).

Upon hearing his answer I was greatly taken aback. The implications of a new generation of secret, highly secure weaponry that exceeds the destructive power of atomic and hydrogen bombs is sobering stuff, to be sure. The conversation ended soon after, as he seemed to be averse to elaborating on what he had told me.

I do not doubt the truth of what this ex-Navy man told me. China Lake has long been a major weapons research and development center for the United States Navy. His account is perfectly consonant with the other information that my research has uncovered.

Deep Beneath White Sands Missile Range, New Mexico

The United States Army's enormous missile test-firing range sprawls for many miles across the desolate deserts, hills and craggy mountain ranges of southern New Mexico. Over the years I have heard bits and pieces of fragmentary information concerning underground facilities and work beneath White Sands. The following account that I received in an e-mail from a distinctly different ex-Navy man is a fine example of the sort of thing I have in mind. He related to me a second-hand account he had received from a water well contractor in southern New Mexico, who had been hired to do some work on the missile range.

> ...he was talking about work he had done at WHITE SANDS MISSILE RANGE.
>
> He was hired to drill 10 or 12 inch dam (sic) holes I think down to 800 to 1000 feet.
>
> He thought nothing of it, well while drilling his bit broke thru (sic) many times. Cold air was rushing up the hole. If I recall correctly he was told not to worry about it.
>
> He later thought to himself maybe he was drilling air shafts?

I think it is entirely possible that the well driller was brought on the base to drill air shafts for deep underground tunnels and structures. My research suggests that New Mexico is prime real estate for clandestine underground facilities - and White Sands Missile Range is right at the top of the list of candidate sites.

Underground Maglev Shuttles?

This is the story that just will not go away. The rumors about super-secret, high speed, deeply buried maglev trains[3] hurtling back and forth, deep beneath the North American continent, turn up again and again. Fellow researcher, Bill Hamilton, relayed the following second-hand account to me of an exchange between a friend and his friend.

> Friend: *So tell me, what's it like to ride in one of those underground shuttles ... ???*
>
> Friend's friend: *VERY DAMN FAST!!! Less than one hour from Cheyanne (sic) Mountain entrance to the pentagon (sic). My calculator would tell me that was over 2000 miles an hour. Odd though, you don't feel the speed, being totally enclosed with no windows. Anyway, there's stuff that's much faster, but not underground.*
>
> Friend: *Wow 2000mph?? Sheesh what a ride. Do you go through some gravitational pull when you take off? Have to wear seatbelts??*
>
> Friend's friend: *There is no more gravitational pull then (sic) when you take off in a bus or train. It doesn't suddenly go 2000 miles an hour. But it does build up faster than you'd think. You don't have to wear seatbelts, but there are warning lights to let you know that you will be leaving, and at the other end, when you are arriving at a slowdown period. They also have overhead speakers that are used for info. You really don't feel much. If you are in a plane, going 600 miles an hour, you get up and walk up and down aisles as if nothing was happening, right? Same on these maglev trains.*

---

[3] Maglev trains, i.e., magnetic levitation trains that hover and ride on a very rapidly moving magnetic field, rather than on rails.

Can these sorts of stories be true? More to the point, is this specific story true? My research indicates that it may be. In my previous book, *Underwater and Underground Bases*,[4] I devoted an entire chapter to documentation from the United States government concerning planned development of a maglev train system. After examining the evidence, I determined that, indeed, it is possible and feasible to build a maglev train system underground. Can I prove that it has been done? No, I cannot. But what I can do – and have done – is demonstrate United States government interest in designing and building a maglev train system. I have also documented extensive underground tunneling activity by a wide variety of government and non-government agencies in recent decades and years. In my previous books, I have also provided firm documentation for a covert operations budget that runs well into the tens of billions of dollars on an annual basis.

So, considering the entire body of available evidence, which includes demonstrable government proclivities for designing and building a maglev train system using the government's *own* documentation, as well as massive sums of untraceable black budget funding, and the underground excavation and tunneling capabilities and technology of American industry – coupled with the persistent stories of ultra-secret, high-speed shuttle trains deep underground – I have reached the point where I believe there are high speed, deep underground shuttle trains operating beneath North America in great secrecy, something that the American people know very little or nothing about.

---

[4] Richard Sauder, *Underwater and Underground Bases* (Kempton, Illinois: Adventures Unlimited Press, 2001).

Nuclear Genie Beneath Burlington, Massachusetts

Some years ago I happened to speak with a man who had a most unusual experience to recount from his youth. Many years ago, upon graduating from college, he took a job with a United States Federal agency which he declined to name.

His work routine was most peculiar: he would enter a perfectly ordinary looking building in the Burlington, Massachusetts area (not far north of Boston) and take a long elevator ride straight down, deep down to a secret, underground laboratory. As he described it to me, the laboratory's work entailed testing radioactive materials and their effects on other materials.

His story has many interesting features: a secret, underground elevator shaft in a normal looking building; the long ride to an underground work area; and the existence of a secret laboratory deep below a major metropolitan area. But perhaps most interesting of all was his description of some of the more senior engineers, technicians and scientists who worked underground. Bizarrely, some of them were missing minor body parts – one man was missing a nose, another a part of a hand, and so forth. So dedicated were they to the nuclear genie that they served that they had sacrificed parts of their bodies, due to radiation burns, in order to advance the nuclear research of the Cold War's military-industrial complex.

This story is as impossible to conclusively prove or disprove as some of the other stories that my sources have recounted to me. I simply do not have the means to go crawling around, hundreds or thousands of feet underground, to verify the salient details of everything I have been told, whether in Massachusetts, New Mexico or California. It's just not possible, whether for me or for anyone else.

Nevertheless, the details of the story are close enough to other little bits and pieces I have gleaned here and there, from a hundred different conversations and untold myriad pieces of documentation that I have seen fit to include it here.

Hollowed Out Norwegian Mountains

The bulk of my research on underground and underwater bases and tunnels has concentrated on American plans, capabilities projects and facilities. The reasons for this focus are two-fold: 1) I am, in the first place, a native-born-and-reared American; and 2) by dint of my long residence in the mainland United States it is more convenient for me to research American plans and activities.

That said, it has nevertheless come to my attention during the course of my research that the Scandinavians are some of the best underground excavators and engineers in the world. The quality and sophistication of their work is very high.

So, it was no great surprise to me to receive a letter from a man who has been inside a mammoth, secret underground base built inside a hollow mountain in Norway. Though the letter is written on United States military stationery, I have elected not to reproduce it as an illustration out of respect for the writer's anonymity. I have, however, seen fit to include a cropped copy of a diagram of the interior of the base, as drawn by the writer, and included in his letter. (See Illustration 2-1.) I have very slightly edited the portion of the letter quoted below to obscure this soldier's identity, rank and military branch.

> My reserve unit was activated just prior to Operation Desert Storm. We are a Cold Weather Unit and as such we were sent to Norway to take part in a NATO War Game "Operation Battle Griffin". It was on this operation that I was sent to do some work inside one of several "Hollowed Out" mountains. At the time, I was a [*here he names his rank-* author's note] and was picked for a work detail along with [*here he identifies the other men-*

author's note] to go to the Mountain. I was rather excited to go on this work detail ... [*excised sentence fragment* – author's note].

Before leaving to go to the mountain we were briefed and inspected which is unusual for a working detail. We were told not to try to see where we were going or which roads we would take. Absolutely no cameras could be taken and nothing considered confidential could leave the mountain. We then left with Norwegian Escorts as they are the only ones who know which roads to take. (According to the briefing the roads crisscrossed in a maze pattern to confuse enemy tank drivers, the Norwegians knew which roads would get us there. We were put in the back of a truck in complete darkness. It was day time but the truck was covered completely.) After driving for a while we came to a stop then heard the sound of Doors opening. We drove into the Mountain and after about a minute or so we stopped. (We were driving at a slow speed at this time.) We got out and looked around. The place was absolutely huge! The smallest area was the entrance which a main battle Tank could easily drive through. The area we were working in was about 300 yards long and maybe 30 yards or so wide the height was about 40 feet or more. It was so big that no one felt cramped or claustaphobic (sic). We were in only one section. I have drawn a diagram of what I saw for you on the back of this letter.

The whole cavity was lined in white plastic type material. This material had various sized zippered doorways, Being curious we opened several of the zippered sections, but all we found was the solid rock of the mountain. [*sentence deleted here* - author's note]. We were inside for several hours stacking gear my unit had used in the operation then left with the same formality we had come.

And here is where my source ends his letter. Brief though his account may be, it nevertheless suffices to convey something of the monstrous size of the facility that he entered. It is noteworthy that the enormous facility that he saw quite possibly was just one portion of a much larger underground base. He additionally alludes to the existence of multiple hollowed out mountains – the underground base he worked in is evidently only one of several.

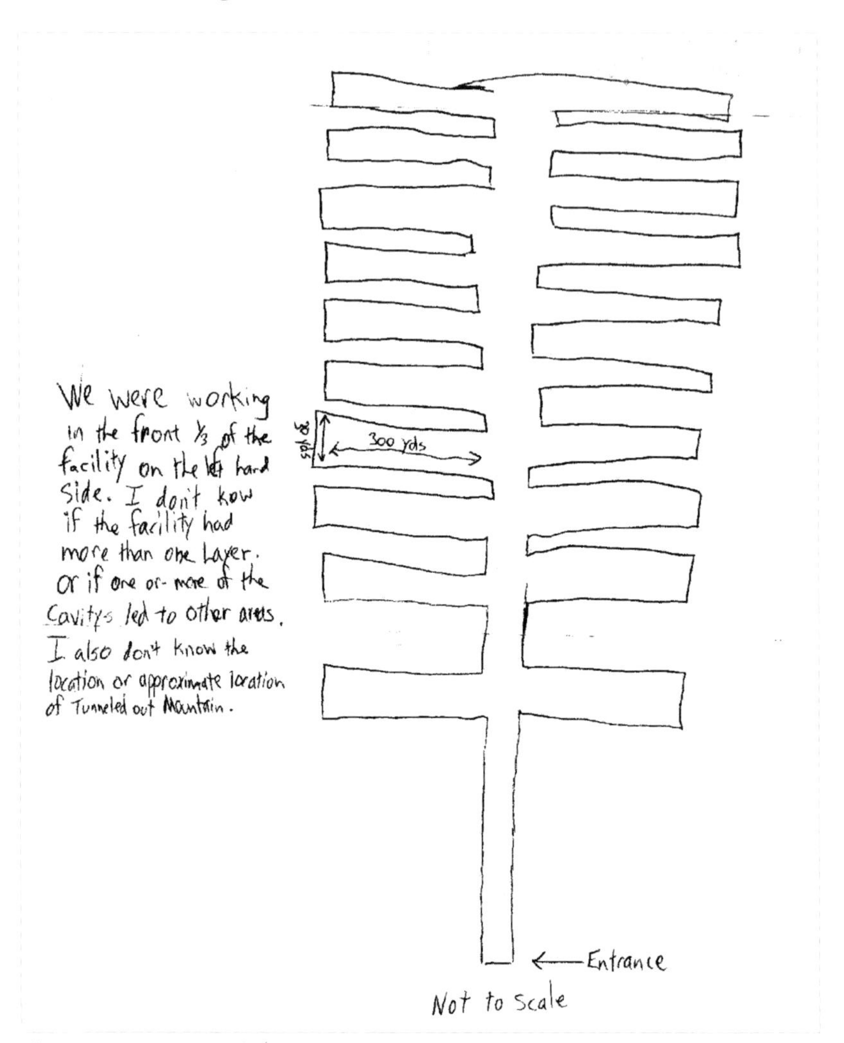

We were working in the front 1/3 of the facility on the left hand side. I don't kow if the facility had more than one Layer. Or if one or more of the Cavitys led to other areas. I also don't know the location or approximate location of Tunneled out Mountain.
80 yds
300 yds
Entrance
Not to Scale

Illustration 2-1. Sketch of secret, Norwegian underground base. Note the huge dimensions – more than 600 yards on the lateral dimension alone. The size of the individual bays is gargantuan. Also, notice that there may possibly be more than one level to this facility. It is impressively large as it is; if there are additional levels or sections then its size is more impressive still. (*Source:* personal letter from member of American military.)

Somewhere Deep Beneath The East Coast of the United States

The following was recounted to me by a man who had worked for the Army Corps of Engineers during the Vietnam War era. As part of his training on American soil, prior to his departure to Southeast Asia, he was briefly sent to tour a couple of underground bases somewhere on the East Coast of the United States.

I asked him where the bases were located and he said, "I don't remember." Of course, this evasive answer is likely to be a lie, presumably due to his security clearances. As you will see, his description of the facilities is sufficiently interesting that it is highly unlikely that anyone, least of all a military man, would forget their locations.

I asked him to tell me a little about the bases and he said that when they sealed up from the outside world, they could do so quite rapidly, within tenths of seconds. Massive, multi-ton reinforced doors like those on bank vaults would slam shut virtually instantaneously. Anyone who had the misfortune to be standing there would be instantly squashed like a tomato. It would all happen much too fast for anyone to react in time to escape instant death.

I asked him about the food supply, and he said that he saw hydroponic gardens in the underground bases, for growing food in the event of prolonged periods of isolation from the topside world.

And that was where our conversation ended. He clammed up and would say little more about the subterranean facilities he had visited.

But what little he did divulge was very revealing. By implication, there are contingency plans for surviving underground for the long term, sealed off from the outside world. And these plans evidently go back at least to the late 1960s to early 1970s.

Secret Bases in the National Forests in California?

I heard something along a similar line in 1995, just after I published my first book, *Underground Bases and Tunnels*. I was discussing my research with a middle-aged woman who had known a young Native American man in Denver, Colorado. She told me that this young man had been quietly approached during the 1980s by officials from the Reagan administration. The officials were inquiring about obtaining non-hybrid, fertile, heirloom seeds from Native American varieties of food crops indigenous to the western states. They wanted these seeds to be taken into and stored in an underground facility then under construction in one of the National Forests in California.

Their reasoning appeared to be that the original Native American food crop varieties had more vigor, and produced better crops under adverse climate conditions, than did the less vigorous, more inbred varieties of seed available on the commercial agricultural market. In the event of a catastrophe that disrupted growing conditions (such as nuclear war? geological upheaval? solar storms? comet and/or asteroid impacts?), they evidently wanted seed varieties that would dependably produce at least some food, so that agriculture could be reestablished, after people left the underground base.

Underwater Base in Gulf of Mexico?

About a half year after the publication of *Underwater and Underground Bases*, I was contacted by a man who said he had knowledge of an unusual underwater project in the Atlantic Ocean. In response to my questioning, he elaborated a little on what he meant by Atlantic Ocean. It turned out that when he said the project was in the Atlantic Ocean he meant that it was located somewhere in the Atlantic Ocean basin. This obviously potentially includes all the bays, gulfs, straits, channels and seas that comprise the greater Atlantic basin, as well as the Atlantic Ocean proper.

Upon further questioning he specified that the project was under the seabed of the Gulf of Mexico, and that Parsons was the contractor. He went on to say that Parsons had purchased some specialty equipment intended for operation 2,800 feet *beneath* the sea floor. This was in about 1997 or so. He thought that the work was somewhere in the eastern part of the Gulf, though he could not be certain.

My source worked for a company that manufactured equipment routinely used in the underground mining and excavation industries. He handled an order from Parsons for an underground project that aroused his curiosity enough to ask a few questions. In response to his questions he was told that the equipment was intended for a deep project beneath the bottom of the Gulf of Mexico. He did not know the project's ultimate purpose. He mentioned too that other companies had also purchased the same equipment from his company for use underwater. The equipment is distinctive enough that it clearly presupposes the presence of live human beings in the places where it is installed.

Now I cannot definitively prove or disprove the presence of a deep facility built by Parsons under the seafloor of the Gulf of Mexico in the closing years of the 20th Century. However, a story like this surely does raise a host of interesting questions. My research has clearly demonstrated that Parsons is one of the premier underground construction firms in the United States, and that it does underground construction and tunneling for a broad range of major corporations and government agencies, including military agencies. Moreover, my research unambiguously shows that the Gulf of Mexico has been the focus of a considerable amount of deep sea drilling by the petroleum industry, as well as by academic and government oceanographers and geologists. Furthermore, there has even been discussion in the open literature of excavating huge caverns beneath the floor of the Gulf of Mexico to store petroleum. It is also the case that the seabed of the Gulf of Mexico is covered with scores of salt domes that would be perfectly suitable for sub-sea excavation and construction of deeply buried, manned, sub-sea-floor bases. So I do not find it hard to believe that Parsons may have simply gone ahead and quietly done what others merely talked about. Such a project would be well within the capabilities of a wealthy, powerful, technologically advanced company such as Parsons.

Indeed, my findings suggest it would not be overly problematic to excavate caverns beneath the Gulf of Mexico and put *something else* in them besides petroleum. That something else could conceivably be almost anything – a secret genetic engineering laboratory, a clandestine prison, a human clone production facility, a highly-secure CIA cocaine warehouse for illegal narcotics in transit between Latin America and the multi-billion dollar North American market, a clandestine nuclear weapons store house, a joint alien-U.S. military base, a super-secure Presidential command center, a super-secure bunker for Parsons corporate executives and directors, an ark populated with a select human population destined to survive possible apocalypse – and let your imagination run with your own possibilities. I have heard two other second-hand anecdotal accounts of secret bases beneath the Gulf of Mexico, one purportedly in the eastern Gulf near Florida, the other in the western Gulf, near Texas. And there is also anecdotal evidence from Puerto Rican coastal waters, in the nearby Caribbean Sea, of U.S. Navy activities that are consistent with the possible construction of one or more sub-seafloor bases there using the Rock-Site methods that I outlined in *Underwater and Underground Bases.*

I am inclined toward the belief that the story my corporate source told me contains a solid core of fact; that there is a clandestine undersea base beneath the Gulf of Mexico, and that Parsons was involved in its construction. Indeed, I am beginning to suspect that there are multiple clandestine, manned, high-tech facilities buried deep beneath the floor of the Gulf of Mexico. If I had to guess, I would say that clandestine U.S. military units are involved there, one or more of the major petroleum companies and other agencies as well.

Such facilities could even house elements from the shadowy world of organized crime and neo-Nazi groups, which have been at the core of the American military-industrial-espionage apparatus beginning with Project Paperclip in the 1940s, and continuing right down to the present day. Thousands of "ex"-Nazis were brought to the United States after WW-II and inserted into positions of influence at the Pentagon, at NASA and in so-called "private" industry. They burrowed away and went to work, networking with each other as they infiltrated the innards of the American military-industrial-espionage complex from the inside out. Over the years, as I have pursued my research, I have run across their trail time and again, in conversation, on the rare occasion in person, in documents in dusty archives. The Nazi imprint on the world did not die with the cessation of open armed conflict between the Axis and Allied powers in 1945. Oh no. The Nazi movement just morphed a little bit and slithered off into another lair across the ocean, there to insinuate itself into the darkest recesses of compartmentalized secrecy, there to lick its wounds and lie in wait for another, better day— its dreams of world conquest held in abeyance for yet a little longer.

Approaching Cataclysm?

So why all the secrecy and clandestine underground and underwater activity? Is Armageddon coming? Is the Apocalypse right around the corner? Do the clandestine movers and shakers inside the military-industrial-espionage complex know something that the rest of us don't? Are we like lemmings unwittingly flocking toward a sheer precipice whose existence we do not even dimly suspect?

At times I wonder. Maybe this is one of the reasons for massive underground bases and secret tunnels. Maybe something BIG is coming up fast?

I actually posed this question to one of my sources and got a wonderfully evasive, sphinx-like answer that frankly raised more questions in my mind than it answered.

As best as I could parse the response, it seems that at least some, and perhaps many, underground and underwater bases and tunnels are preparations for the following *possibilities* in the relatively near term (say within the time-frame of the next several years to next few decades):

- potential nuclear war
- change in physics of rotation of the Earth, i.e., a pole shift, or axis relocation
- reversal of Earth's magnetic field
- entering a galactic dust cloud
- asteroid and/or comet strikes
- massive, sudden climate change

- social, political and economic chaos resulting from any of the preceding
- a combination of any one or more of the above.

In other words, very dramatic events may be just around the corner. It seems as if those in the know are making massive, extremely expensive preparations for something.

Conclusion: Import of the Anecdotal Stories

Perhaps the most interesting aspect of the preceding accounts is that most of them were recounted directly to me by people who worked for the military-industrial complex, in some cases in the uniformed military services themselves.

This is a recurring theme in my research. Time and again I have been approached by ex-military personnel, and, I believe, even active-duty military personnel in some cases. Why should this be the case? I believe it is because some people who are involved in the activities of the United States military see things that bother them, or perhaps are asked to carry out duties that trouble them. Some of these people then leak information to a researcher such as myself, in hopes that some of what they have seen, or have been asked to do, will trickle out into the public domain, for the greater edification of the general population.

I can only speculate as to their personal motivations, but I think their concerns must be personal ethical misgivings, in some cases, deep political unease at the machinations of secret, unconstitutional government, and even profound strategic military misgivings in others.

For example, it may be technologically possible to design, build and deploy continent-busting weaponry that far exceeds the destructive power of nuclear weaponry, but is it morally advisable to do so? Is it militarily advisable to do so? Under what conceivable set of circumstances would a military command structure literally unleash hell on Earth, such that entire continents, or regions of continents were laid waste? What sane person would set in motion such a series of events? What sane person would care to inhabit a planet that had been reduced to a smoldering, radioactive cinder? Under what fiendish mind-set could such an outcome be considered a "victory"?

Or again, if the underground and underwater infrastructure of secret bases and tunnels has possibly grown so extensive, so sophisticated and so out-of-the-oversight of the constitutionally mandated political checks and balances prescribed by open, civil government, then it may be that the well-being of the American people, and perhaps many others as well, is profoundly threatened by a stealthy, massively funded, high-tech, underground and/or undersea power about which they remain blissfully unaware, the nefarious intentions and plans of which run directly counter to their best interests.

Indeed, if I had to guess, I would surmise that more than a few of my sphinx-like sources have concerns that run very much along these lines. They see things they do not approve of, but they are severely hemmed in and constrained by rigorous security clearances that carry severe penalties for violations. What would you do, in their place, if you saw something with which you profoundly disagreed, but could not speak out openly, for fear of being sentenced to many years in prison, and/or very heavily fined? Or perhaps even targeted for personal liquidation, i.e., assassination, for speaking publicly? You just might very quietly and anonymously leak a bit of information to an investigative author in hopes that it would find its way into his next book, or magazine article, or radio interview, and thereby find its way out into the world and the light of day where thousands, eventually millions, of other people could read it, hear it, and ponder its significance.

Chapter 3:

Secret Tunnels Beneath Washington, D.C.

From the very beginnings of my underground base and tunnel research in the early 1990s, it seemed to me that there must be secret tunnels beneath Washington, D.C. Of course, I know about the metro system, much of which runs in tunnels that lie underground beneath Washington, D.C. and neighboring counties in the nearby suburban areas of Maryland and Virginia.[1] I have ridden the metro trains in Washington, D.C. many times. And I've always felt that there must be something more beneath Washington, D.C. than just the metro system – more tunnels, much more secret, and much deeper underground. After all, the seat of the Federal government is in Washington, D.C. and the surrounding areas of Maryland and Virginia. Wouldn't it make sense to have multiple levels of security? Physical security? Deeply buried protection, well away from prying eyes and ears, and safely beyond the electronic probing of spy satellites and the destructive blast effects of a possible nuclear strike in the event of a nuclear war?

---

[1] Washington, D.C. Metro System Map, http://www.wmata.com/metrorail/systemmap.cfm, 2004.

Moreover, given the documentation I have found about underground facilities in the mid-Atlantic region, including in the Washington, D.C. region itself, it seems like it would make sense to also have a secret tunnel system connecting these sensitive installations with each other. My research has clearly shown that construction of a regional – or national – secret tunnel system is technically feasible.

To be sure, the idea of deeply buried tunnels underlying a major city is nothing new. Many cities and towns, large and small, all over the world, have all kinds of tunnels running beneath them. I run into this sort of fascinating information all the time in my research. Indeed, anyone who has ever traveled on the Tube, deep under London, or taken a ride on the New York Subway realizes that all major cities have many miles of underground tunnels crisscrossing the urban landscape. Besides metro tunnels, as in Paris, London, New York and Washington, D.C., there are also highway tunnels, passenger train and freight tunnels, sewers, aqueducts, steam tunnels, conduits for electrical and fiber optic cables, tunnels for pedestrians and more. The subterranean depths beneath many a major city, and even smaller cities, can be a veritable labyrinth of passageways.[2]

As an example, take a look at the cutaway sketch of Piccadilly Circus in London (Illustration 3-1), rendered as it appeared circa the early 20th Century.

I have visited London myself, and even strolled through Piccadilly Circus without being aware of the anthill-like maze that lay beneath my feet. Notice that there are four train tunnels buried beneath Piccadilly Circus, several large escalators and staircases,

---

[2] Alex Marshall, *Beneath the Metropolis: The Secret Lives of Cities* (New York: Carroll & Graf Publishers, 2006). This is an excellent introduction to the subterranean labyrinth that lies under every major, modern city.

multiple pedestrian passageways, a booking hall, a machinery floor, shops and vendors' stalls, and the space to simultaneously accommodate many hundreds of people in transit. Amazingly, the entire warren of booking hall, tunnels and passages beneath Piccadilly Circus was excavated through one 18-foot diameter shaft![3]

Illustration 3-1: Subterranean maze beneath Piccadilly Circus in London, circa early 20th Century. *Source*: London Transport Museum.

[3] Patrick Beaver, *A History of Tunnels* (Secaucus, New Jersey: The Citadel Press, 1972).

Williamson Tunnels Under Liverpool, U.K.

An even more intriguing example of a subterranean labyrinth is the so-called Williamson Tunnels that lie beneath Liverpool, U.K. Joseph Williamson, a wealthy merchant and philanthropist in early-19th Century Liverpool, kept large gangs of men, numbering in the thousands in the aggregate, busy over a period of about 35 years tunneling beneath Liverpool's Edge Hill district, making one tunnel after another. No one knows the full extent of the original tunnel labyrinth, but it may extend for miles; perhaps with tunnels that reach even as far as the center of Liverpool. In any event, the work stopped upon Williamson's death in 1840 and the tunnels have been largely abandoned since that time.[4] I find this example illustrative, because it shows that just one wealthy man, employing gangs of laborers using the pick and shovel technology of the early 19th Century, could create a subterranean labyrinth of considerable extent under a major metropolitan area. Imagine how much more today's technology could accomplish using modern technology and backed by the power of major corporations or government agencies.

Salt Mine Below Detroit, Michigan

There is another interesting case I want to mention, due to its great depth. Comparatively few people outside of Detroit know that there is a huge salt mine with 50 miles of roads tunneling through the subterranean depths more than 1,100 feet below Detroit, Michigan. The salt mining began in 1896 when a 1,100 foot shaft was sunk. Mining continued until 1983 when market forces brought operations to a halt; however, in recent years the mine has opened

---

[4] For more information see the "Friends of Williamson's Tunnels" website at http://www.williamsontunnels.com, 2004 and also the "Williamson Tunnels Heritage Centre" website at http://www.williamsontunnels.co.uk, 2004.

again.[5] How many people who visit Detroit are aware that more than 1,000 feet below their feet monstrously huge, 40-ton trucks with seven-foot diameter wheels prowl along 50 miles of roads carved out of solid rock salt?

SubTropolis Under Kansas City, Missouri

There are so many examples of major cities with unusual subterranean levels that could be mentioned, and I will turn back to Washington, D.C. in a moment, but I would be remiss if I did not also briefly mention SubTropolis, in Kansas City, Missouri. There are thick beds of limestone in the Kansas City area that have been extensively mined out in years past. These miles of mined out tunnels and caverns are owned by Lamar Hunt and his family, and have been turned into the world's largest underground business complex. Hunt Midwest SubTropolis, as it is called, houses more than 50 businesses, including warehousing, cold storage, light manufacturing and office space. Considered a foreign-trade zone, the SubTropolis has 4.3 million square feet of building space, and more than 20 million square feet of developed space. More than 1,300 people work underground in the SubTropolis. Its miles of brightly lit roadways accommodate cars and large, over-the-road, tractor-trailer trucks.[6] How many visitors to the Kansas City area suspect the presence of miles of underground, brightly lit streets and a subterranean business and light industrial, foreign-trade zone? Although to my knowledge there is nothing covert about SubTropolis, I knew nothing about it

[5] Patricia Zacharias, "The Ghostly Salt City Beneath Detroit," *The Detroit News*, http://www.detnews.com/history/salt/salt.htm, 2002.

[6] For more information see: "SubTropolis, Kansas City's Basement," http://www.bizsites.com/1999/insites5.html, 2002; and also "What is SubTropolis?," http://huntmidwest.com/Subtropolis/WhatIsSubtropolis.shtml, 2004.

before beginning my research, even though I had personally spent some time in the Kansas City area in the early 1980s.

Secret Tunnels Underlie Tokyo

Then there is the curious case of Shun Akiba's research in Tokyo, Japan. Shun Akiba is a former broadcast journalist, who worked as a TV reporter covering international conflicts for Asahi TV. In 1996 he left television to become a freelance writer. But his career took a completely unexpected turn when he stumbled across information that suggested there was an extensive system of unknown train tunnels that had been secretly constructed beneath Tokyo. As he examined construction records and tunnel maps, both old and new, he came to the startling realization that there were many tunnels and unexplained underground spaces and levels that were not accounted for by the official records. He believes that there may be 2,000 km of tunnels below Tokyo, far more than the 250 km of publicly known subway tunnels.

Shun Akiba has compiled his findings and conclusions in a book entitled, *Imperial City Tokyo: Secret of a Hidden Underground Network* (Yosensha, 2002). Notwithstanding the dramatic import of his research the mainstream news media have ignored Shun's findings and his work suffers from relative obscurity. Interestingly, his book has gone to five editions,[7] though it is not available in English, to my knowledge.

I could go on citing more examples and belabor the point, but by now you should have the idea. Lots of cities, including major world capitals, have all sorts of subterranean levels that crisscross the depths far below; in some cases secretly, as in the Tokyo example, and in

---

[7] Angela Jeffs, "Seven Riddles Suggest a Secret City beneath Tokyo," *The Japan Times*, 1 March 2003, http://www.japantimes.co.jp/cgi-bin/getarticle.pl5 ?fl20030301a1.htm, 2003.

some cases very deep underground, indeed, as in the case of the Detroit salt mine tunnels. Not surprisingly, my research suggests that Washington, D.C. is no exception.

Washington, D.C. Metro

Millions of people who have lived, worked, or traveled in the Washington, D.C. area have ridden the metro. Crowds of tourists, government workers, school students and others ride the trains beneath the nation's Capital every day. It is a well-run system and operates relatively smoothly, especially for a big city transit system.

But it is important to realize that the metro system is not the beginning and end of underground tunnels in the D.C. area. Oh no, not at all. Planning for the tunnels began in the 1950s, though construction did not begin until 1969,[8] and continued thereafter for about 30 years, until the system was finally completed. But the serious tunneling under Washington, D.C. had begun long before. And even in the 1960s, when the metro planning and construction were gearing up, there were other plans for other underground work, much more secret, and far, far deeper, as I shall shortly show.

The Mysterious E-Mail

One of the things that happens to authors who write the sorts of books I have written is that people write in with information and tips – they send letters to me, courtesy of my publisher, or send me e-mails. And sometimes what they have to say is highly interesting. Let me give you an example. In late summer 2004, I received an e-mail from a source who alleged to be a federal employee in the Washington, D.C. area. Here is what my source said:

---

[8] *Building the Washington Metro*, http://chnm.gmu.edu/metro/, 2004.

> ... in washington, dc, there is a huge metrorail underground complex. at one of the metrorail stations in the city, there is a door a little ways down a tunnel that leads to an intermediate terrace where another door will provide access to either a ladder or elevator that will take people to the underground tunnels that connect the whitehouse with other places that get used in time of war when a shadow government needs to operate.

That's it. That's the story, written in all lower-case type as you see above. The question is whether what the author of the e-mail alleges about secret underground tunnels beneath Washington, D.C. is substantially true. I am inclined to think that it is. I will lay out for you the evidence suggesting that my anonymous source is on to something, that there is indeed a secret tunnel system under Washington, D.C.

Hogs In The Tunnel!

A few years back, a mysterious, little booklet with the words, *Hogs In The Tunnel!*, emblazoned on its cover crossed my path. As I recall, I had recently published my first book and somehow let the booklet slip away. Happily, when I spoke at Steve Bassett's first X-Conference in Bethesda, Maryland in April of 2004, one of the conference attendees mentioned that he had the booklet and kindly sent it to me.

And a most peculiar thing it is! A delightful little treasure, a genuine subterranean artifact written by Washington, D.C. journalist, John Elvin. *Hogs In The Tunnel!* is a byproduct of his personal curiosity about the nether realms below Washington, D.C. and his wealth of insider's knowledge accumulated from years of reporting for the *Washington Times* newspaper in his *Inside The Beltway* investigative column, and for his muckraking newsletter, *Political Dynamite.*

In a nutshell, Elvin reveals that there is a literal warren of tunnels beneath Washington, D.C. The *Hogs In The Tunnel!* phrase on the cover refers to the purported cry that the Secret Service detail for Bill and Hillary Clinton would allegedly holler whenever the Clintons decided to secretly step out on the town via the tunnel network that radiates out from the lower levels below the White House. Why the cry, "Hogs in the tunnel!"? Simple: the Clintons moved to Washington, D.C. from Arkansas, Bill Clinton's home state. As it happens, the mascot for the University of Arkansas, where President Clinton formerly taught law before becoming governor of Arkansas, is the razorback hog. Consequently the hog moniker was applied to the Clintons, as transplanted Arkansans, whenever they entered the tunnels. Additionally, the deceased gonzo journalist, Hunter Thompson, supposedly used the phrase.

According to Elvin, his research and insider sources revealed that beneath Washington, D.C. lies a "serpentine network of secret underground passageways ... [that] looks like miles and miles of huge intestines, wrapped back and forth, bigger tubes and smaller corridors, as if the earth had its very own innards."[9]

Elvin dates the beginnings of the secret tunnel system below Washington, D.C. to the very early days of the Republic, when a British attack was feared – and justifiably so, as events in the early 19th Century showed. One of the early tunnels ran from a secret stairway in the Octagon House at 18th and New York Avenues and ended behind a brick fireplace in the White House, at 1600 Pennsylvania Avenue, with a side tunnel that burrowed off to the Potomac River, for use as emergency escape.[10] Other reported tunnels run

[9] John Elvin, *Code Words: "Hogs In The Tunnel!": An Underground Classic* (Atlanta, Georgia: Soundview Publications, 1995), pp. 6-7.

[10] Ibid., 11.

from the State Department to the White House, and stretch the length of Pennsylvania Avenue. Some of these tunnels are said to be impressively large, and able to accommodate vehicular traffic. Elvin says that Georgetown, the exclusive, upscale residential district on the west side of Washington, D.C., is a "beehive of tunnels,"[11] and he describes the many tunnels beneath the Capitol and its environs, as well as the network said to radiate out from beneath the Treasury Building, and thousands of miles of other tunnels of every description, some damp and slimy and clammy, others well-lit and monitored by security patrols.[12]

Elvin alludes to the elaborate Presidential facility buried deep beneath the White House, and to the large tunnel that runs the length of Pennsylvania Avenue and muses as to whether the White House might not actually be a sort of Potemkin Village, i.e., an elaborate, false facade to conceal the fact that the President and his family reside elsewhere, perhaps coming and going from the White House via the many secret, underground tunnels that underlie Washington, D.C.[13]

Pentagon Pneumatic Tube To Dulles Airport

One of the most intriguing stories that Elvin reports has to do with something that he says he never reported elsewhere: "a huge pneumatic tube that can shoot cargo, or a person, in a pellet right down the middle of the access road to Dulles Airport in outlying Virginia." John Elvin talked to one of the men who worked on the tunnel, a man he has known for years, and says that he has no reason

---

[11] Ibid, 14-15.

[12] Ibid., 17-25.

[13] Ibid., 28-33.

to disbelieve him. The tunnel purportedly begins at the Pentagon.[14] I personally find this story interesting because for the last several years I have gleaned fragmentary bits and pieces of information, here and there, none of it conclusive enough to say anything with great certainty, that hint of a secret tunnel system radiating outward from the Washington, D.C. area to other locations in the mid-Atlantic region. This information from John Elvin lends credence to such a scenario. Additionally, my own research reveals that the federal government mounted a serious research effort into designing and building deep underground pneumatic tunnels as a transportation system, back in the 1960s and 1970s. Have such underground, pneumatic tube tunnels really been built, albeit in great secrecy, via the black budget, for the clandestine use of the federal "alphabet soup" agencies? I am inclined to think they have.

Presidential 'Escape Train' Under New York Hotel

To the uninitiated, this business of secret, Washington, D.C. tunnels for bigwig politicians and other clandestine purposes may seem a bit much to swallow. But consider this fact: in September 2003, when President George W. Bush spent a couple of days at the posh Waldorf-Astoria Hotel in New York City, a special train was kept idling underground beneath the hotel for his use as an escape in an emergency. The Secret Service had the train stationed "at an abandoned platform" known as Track 61; the President would have accessed it via an "underground passage from inside the landmark hotel and by a freight elevator that descends from a brass-sheathed door next to the hotel's parking garage..."[15]

---

[14] Ibid., 34-35.

[15] "Bush 'escape train' under N.Y. hotel," MSNBC, 26 September 2003, http://www.msnbc.com/news/972285.asp, 2003; also http://msnbc.msn.com/id/3087301/, 2005.

If this much is admitted in the newspapers and on the Internet, it is conceivable that still more secret options were in play. To wit, perhaps this level of security was itself a screen for a still more secure and even more secret escape option? My research does not exclude the possibility that Manhattan could well have secret tunnels that the public knows nothing about. And if the President is prepared to use a secret tunnel beneath the Waldorf-Astoria Hotel in New York City, then why not in Washington, D.C., too?

A Not-So-Secret Tunnel

A more recent example of a clandestine tunnel beneath Washington, D.C. came to light in 2001 in connection with the Robert Hanssen spy scandal. Hanssen was the FBI counter-espionage agent who was arrested and jailed for spying for the Russians. News reports in March of 2001 revealed that Hanssen evidently tipped off the Soviets that during the 1980s the Federal Bureau of Investigation (FBI) and National Security Agency (NSA) had secretly dug a surveillance tunnel beneath the Soviet Embassy in Washington, D.C. The project reportedly cost several hundred million dollars.

When asked about the clandestine project on the CBS News *Face the Nation* television news show, Vice President Dick Cheney commented: "If it were true such a tunnel were dug, I couldn't talk about it anyway."[16] As we shall shortly see, this is an interesting response by Dick Cheney, indeed!

What's Under The Vice-Presidential Mansion?

Just a little over one year after giving the evasive response to CBS News about the reported secret tunnel under the Russian Embassy,

[16] "A Not-So-Secret Tunnel," CBS News, 5 March 2001, http://www.cbsnews. com/stories/2001/03/04/national/main276193.shtml, 2002.

Dick Cheney's neighbors began to complain about a construction project at the Vice President's residence that created repeated, huge blasts that were rocking their neighborhood – and the explosions emanated from Dick Cheney's own house! Now, it should be said that the Vice-President of the United States lives in a mansion on the grounds of the U.S. Naval Observatory in northwest Washington, D.C., just off of Massachusetts Avenue, and the location is surrounded by expensive homes in a wealthy section of town.

So you can imagine the concerns that residents of the neighborhood had when they began to hear loud explosions 2 or 3 times a day, early in the morning, as well as late at night. When neighbors complained about the continuing blasting, the Naval Observatory Superintendent, David W. Gillard, sent out a letter, in which he said:

> Due to its sensitive nature in support of national security and homeland defense, project specification is classified and cannot be released. In addition, please understand we are severely constrained by operation requirements to perform this project on a highly accelerated schedule...[17]

Gillard went on to say that the blasting could last for another eight months, and also said that the Navy had tried to create less noise on the construction project (whatever it was) by "silencing backup alerts on trucks and removing most diesel-powered electric generators from the construction site." A Navy spokeswoman further characterized the work as "infrastructure and utility upgrades." Local residents speculated that the work may involve underground excavation or tunneling.[18] I am inclined to agree with this view.

---

[17] David Nakamura, "Cheney's Home Sending Bad Vibrations," *Washington Post*, 9 December 2002, http://www.washingtonpost.com/wp-dyn/articles/A24386-2002Dec7.html, 2002.

[18] Ibid., and "Secret blasts rattle Cheney's neighbors," CNN, 8 December 2002, http://www.cnn.com/2002/US/South/12/08/noisy.neighbor.ap/index.html, 2002.

Put this in perspective: when is the last time *you* tried setting off high explosives at *your* house? Especially in these troubling times of *Homeland Security*? For most people, that sort of activity would probably earn a swift trip to a prison cell – but not for the Vice President of the United States, whose house needed hundreds of explosive blasts over a period of many months!

Now, any thinking person can plainly see that while all this blasting was going on, Vice-President Cheney must certainly have been somewhere else. Perhaps down in the labyrinth of tunnels that underlies Washington, D.C., since presumably Cheney – the consummate Washington, D.C. insider – would already have known about the tunnels and, more than likely, would have regularly been using them himself.

But the highly unusual level of heavy construction activity at the Vice Presidential mansion clearly indicates a more serious kind of undertaking – perhaps going deeper, much deeper? In this regard I cannot help but think back to the Golden Eagle Award that Dick Cheney received in 1996 from the Academy of Fellows of the Society of American Military Engineers (SAME). Two Golden Eagle Awards are given each year to "engineers who have made singularly distinctive contributions to the profession of military engineering and to America's defense establishment." In 1996, Cheney shared the honors with Steve Greenfield, the Chairman Emeritus of the Board of Parsons Brinckerhoff, Inc., perhaps the premiere underground construction and tunneling company in the United States.[19] The interesting thing about this engineering honor is that while Cheney is a past Secretary of Defense, he is not an engineer. His official biography plainly states that his academic training is in political

[19] "Notes on the Firm," http://www.pbworld.com/PbinPrint/Notes/Spring97/8-97NOF.htm, 2002.

science, and his adult career has been in the political arena.[20] It is also interesting that Cheney received this honor along with the Chairman Emeritus of the Board of Parsons Brinckerhoff, Inc.

Did Cheney and Greenfield perhaps work together on a major, clandestine, underground project, with Greenfield handling the hard engineering side of the project, while Cheney pulled the levers of power in the smoky, back rooms of the black budget world that clandestine movers and shakers navigate to get big, secret things done? I don't know, but there are 3.3 *trillions* of dollars (that would be 3.3 thousand billions) unaccounted for in the American black budget,[21] and Parsons Brinckerhoff, Inc. has popped up in my research more than once, as has Dick Cheney. When a non-engineer such as Dick Cheney receives a prestigious engineering award, that is an interesting turn of events and begs the question: for what?

Navy Keeps A Secret In Plain Sight

But there's more. Just a few years ago a noteworthy little item appeared in the *Washington Post* newspaper. The *Post* is supposedly the newspaper of record for official Washington, D.C., so I suppose the information in the article is about the closest we'll come to an official announcement of strange goings-on down by the Potomac River, in the southeast part of town.

The title of the article was, "Navy Keeps A Secret in Plain Sight," and the sub-title was, "Hush-Hush Project Underway by Potomac." What it all boils down to is that the U.S. Navy had a secret construction project going on in Washington, D.C. and it wouldn't tell

---

[20] "Vice President Dick Cheney," http://usinfo.state.gov/topical/transition/cheney.htm, 2002.

[21] Catherine Austin Fitts, "The $64 Question: What's Up With The Black Budget?," http://www.scoop.co.nz/mason/stories/HL0209/S00126.htm, 2005.

anyone what it was all about. Well, not quite, seeing as how the story was splashed across the pages of the *Washington Post.*

And the article did, in fact, reveal some details:[22] 1) The entire, mysterious construction site was fenced off, down by the Jefferson Memorial, on national parkland. 2) A Navy spokesman said that, "As a matter of policy, we can't go into the particulars." 3) The normal, multi-agency review for construction on national parkland was ignored on this project, and officials who were briefed about the project were sworn to secrecy. 4) The article referred to "hangar-like structures, which cover an excavation area..." and alluded to an expected project duration of four years. 5) No paper trail was produced for this "security-sensitive project". 6) National Park employees were reportedly told that the secret construction was "utility work," while the Navy said that the work was a "utility assessment and upgrade." 7) The Acting Secretary of the U.S. Commission of Fine Arts, which has oversight of the Potomac River parkland in question, said that it was "illegal" for him to discuss what was happening. Indeed, the Navy began the construction without even informing the National Park Service, which owns the land. 8) "...(D)igging continues...after more than 300 days" – that's right, "digging" – and one of the two companies involved was Kiewit.

As it happens, Kiewit is a firm that I know well from my own research, since it is a known constructor of underground military facilities, and has also excavated lengthy, deep tunnels for a variety of government agencies.[23] Therefore, even if the U.S. Navy ultimately wrapped up its project in a few years as it said it would, and left

---

[22] Spencer S. Hsu, "Navy Keeps A Secret in Plain Sight," *Washington Post*, 26 November 2004, http://www.washingtonpost.com/wp-dyn/articles/A13265-2004Nov25.html, 2004.

[23] Kiewit Underground, http://www.kiewit.com/project/underground.asp, 2005; and Kiewit Defense, http://www.kiewit.com/project/defense.asp, 2005.

behind nothing more than a "small equipment shed" and "agreeable landscaping" (according to the article), reasonable doubt will always remain in my mind as to what the Navy has actually done. My research clearly shows that a relatively unobtrusive building can serve as an entry point for a major underground working. The building can be anything at all – the more unlikely the better, if the purpose is to misdirect people's attention. Of course, these types of questions would not arise at all, but for the Navy's high handed, overweening secrecy, and its total disregard for established legal procedures long in place. In other words, by its own behavior the Navy has indicated that *whatever it did cannot possibly be business as usual.* The presence of a known, deep tunneling company on the project (Kiewit), along with references to more than 300 days of digging, inevitably leads to the conclusion that whatever happened at the site occurred deep underground.

Meanwhile – Back at Dick Cheney's House....

All of which reminds me of the loud blasting and heavy construction at Dick Cheney's mansion in 2002, which was also a hush-hush, secret U.S. Navy project that could not be discussed in any detail. Remember that, according to a Navy spokeswoman, the project at Cheney's house involved "infrastructure and utility upgrades." This is reminiscent of the project along the Potomac, which the Navy also said involved "utility assessment and upgrade." The language the Navy spokespersons used to describe the two projects is very similar – and in fact, the projects are very close together in time, only a couple of years separate them, and from a geographic point of view, are only several miles apart as the crow flies. Could the two "secret" U.S. Navy projects be related? I don't see why not. The two construction sites could conceivably be two phases of the same project, the bulk of which lies deep underground. I don't positively know this to be a fact, mind you, but the Navy's air of secrecy, and the type of work involved, as well as the presence of Kiewit on the job site by the Potomac, raise the issue. And there is another germane factor, which I have thus far not mentioned, which brings the discussion to a much deeper level still.

The LBJ Presidential Library Documents

A few years ago I was browsing a website on the Cold War[24] when I noticed a reference to some declassified materials on a so-called Deep Underground Command Center, to be built far below the Washington, D.C. area, at the height of the Cold War. I was intrigued, so I corresponded with the website's owner to ascertain if the documents were genuine. He assured me that they were and told me exactly where to find them, at the Lyndon Baines Johnson Presidential Library, on the campus of the University of Texas, in Austin. So I traveled to Austin and found the LBJ Presidential Library, and the documents, exactly as he had indicated. They are most revealing and show that already, back in the early 1960s, serious plans were being made at the very highest levels of the United States government to build a very deeply buried command center, approximately 3,500 feet beneath the Washington, D.C. area.[25]

Interestingly, the first of these high-level memoranda begin in November 1963, just a couple of weeks before President John

---

[24] "A Secret Landscape: The Cold War Infrastructure of the Nation's Capital Region," http://coldwar-c4i.net/index.html, 2005; and "The Deep Underground Command Center (proposed)," http://coldwar-c4i.net/DUCC/index.html, 2005.

[25] These documents are located in the Lyndon Baines Johnson Presidential Library, on the campus of the University of Texas, in Austin, Texas. The records are, respectively, in the National Security File, Subject File, Deep Underground Command Structure, Box 8, and National Security File, Files of Spurgeon Keeny, Deep Underground Command Center, Box 3.

Kennedy was assassinated. I have no idea if these plans have any relationship to the assassination of the President; however, the paper trail shows that right about the time that Kennedy was murdered, in late November 1963, he was scheduled for a series of briefings on a number of matters, including the question of the Deep Underground Command Center. After the Kennedy assassination, this TOP SECRET policy matter passed on to the Lyndon Baines Johnson administration.

Deep Underground Command Center (DUCC)

The first TOP SECRET memorandum issued from the office of the Secretary of Defense, Robert McNamara, on 7 November 1963, just two weeks before the Kennedy murder in Texas. In the memo, McNamara set out a number of topics that he wanted to discuss with the recipients in the following couple of weeks, including the subject of the National Underground Command Center. The recipients of the memo were McGeorge Bundy, Kermit Gordon and Dr. Jerome Wiesner. McGeorge Bundy was the President's Special Assistant for National Security Affairs (and stayed on in the same capacity for Lyndon Johnson after the assassination). Kermit Gordon was Director of the Bureau of the Budget and later became the President of the Brookings Institution in 1967. Dr Jerome Wiesner was Science Adviser to President Kennedy, and later became President of the Massachusetts Institute of Technology. McNamara indicated that they would likely meet with the President sometime during the last week of November to discuss the topics in the memorandum. Unfortunately, Kennedy was dead by then, having been assassinated in Dallas, Texas.

The second TOP SECRET memorandum, seven pages in length, was also issued on 7 November 1963. It is a Memorandum For The

President concerning a proposed Deep Underground National Command Center, to be placed 3,500 feet underground in the Washington area. According to the information provided in the memorandum, the proposed facility was intended to survive multiple direct hits of 200 to 300 megaton warheads impacting at ground level, or 100 megaton warheads that penetrate 70-100 feet below the surface before detonating. Clearly the concern of these planners was the waging and survival of nuclear war. Red China and Russia were both explicitly mentioned as potential adversaries of the United States in a shooting nuclear war. Two alternatives were presented for the President to consider: 1) An austere facility with 10,000 total square feet and 5,000 operating square feet, accommodating 50 people at a depth of 3,500 feet. The cost of construction was estimated at $110 million. 2) A moderate sized facility with 100,000 total square feet and 50,000 operating square feet, accommodating 300 people at a depth of 3,500 feet, and costing $310 million.

The memorandum explained that two years would be needed to sink a vertical shaft to the required depth. It further specified that the "operational capsule" of the facility would be near the Pentagon and that an access elevator would descend directly from the Pentagon to the facility. The memo also mentioned elevator shafts below the State Department and White House that would descend to the 3,500 foot level, with high speed, horizontal tunnel transport to the deeply buried main facility. Key personnel could leave their offices in these major office buildings and descend "undetected" to the 3,500 foot depth within 10 minutes. Within 15 minutes they would be in the Deep Underground Command Center (DUCC). Remember that these projections were made using 1960s-era technology.

Reading further, the memorandum stated that the JCS (Joint Chiefs of Staff) felt that the austere facility was too small, and solicited the views of the President on the DUCC concept.

Importantly, the document forthrightly stated: "There is little argument that construction of a DUCC is technically feasible..." And it observed: "The nature of the DUCC, regardless of the initial size selected, lends itself to later expansion." The recommendation of the report was that work on an austere DUCC be approved and that construction should begin using fiscal year 1965 funding, with the option to expand the facility to the 300-man moderate size after construction had begun. The author, who was not directly named, but evidently was Robert McNamara, said that: "...I am convinced that a DUCC of at least the size of the austere proposal is required." The last sentence of the memorandum stated: "The Secretary of State has seen this paper and concurs."

These documents clearly indicate that initial planning for an extremely deeply buried command center below Washington, D.C. had begun at a TOP SECRET level in the United States government, as of two weeks immediately prior to the Kennedy assassination. The papers nowhere indicate what Kennedy himself thought of this policy. However, other declassified memoranda written in the months immediately following Kennedy's death, in the new Johnson administration, indicate that the idea remained very much alive and under active consideration.

~~TOP SECRET~~ 5

THE SECRETARY OF DEFENSE
WASHINGTON

7 November 1963

MEMORANDUM FOR: Mr. McGeorge Bundy
Mr. Kermit Gordon
Dr. Jerome Wiesner

In each of the past two years we have reviewed with you the major issues affecting the Defense Budget before presenting the Budget to the President. We should like to follow the same procedure this year. As a basis for the review, I shall send to you copies of "Draft Memoranda to the President" covering the following subjects:

1. Strategic Nuclear Forces
2. Continental Air and Missile Defense
3. Army and Marine Ground Forces
4. Land Based Tactical Air Forces
5. Attack Carrier Forces
6. Anti-submarine Warfare Forces
7. Airlift and Sealift Forces
8. The National Underground Command Center
9. The Research and Development Program.

I propose that we meet together on Friday, November 15, at 1:30 in my office to discuss items 4, 5, 7, 8 and 9, and on Friday, November 22, at 1:30 to discuss items 1, 2, 3 and 6. Following such meetings, I hope we will be prepared to discuss the issues with the President at his convenience on or after November 25. He may wish to meet for this purpose on Friday, November 29, the day after Thanksgiving.

Attached are copies of the memoranda on the National Underground Command Center and the Research and Development Budget. Memoranda relating to items 4, 5 and 7 will be sent to your office later today or early tomorrow.

Please ask your secretary to inform my office if these arrangements are satisfactory to you.

Robert S. McNamara
Robert S. McNamara

cc: DepSecDef
DDR&E
ASD(Comp)

S&T TOP SECRET Cont. No. 345

~~TOP SECRET~~

Illustration 3-2: Robert S. McNamara memorandum, 7 November 1963, in which John F. Kennedy's Secretary of Defense discusses an upcoming meeting in which one of the items on the agenda is a planned, deeply buried underground base beneath Washington, D.C.. The time frame is just a couple of weeks before the president was assassinated in Dallas, Texas. *Source:* Lyndon Baines Johnson Presidential Library, Austin, Texas.

FINAL DRAFT
November 7, 1963

~~TOP SECRET~~

SANITIZED
E.O. 12958, Sec. 3.6
NLJ 97-83
By ___, NARA Date 2-27-98

MEMORANDUM FOR THE PRESIDENT

SUBJECT: National Deep Underground Command Center as a Key FY 1965 Budget Consideration

A continuing examination of the problems associated with providing an adequate national command and control structure to meet contingencies that might occur in the 1970-1975 time period prompts serious consideration of the construction of a Deep Underground Command Center (DUCC) in the Washington area.

As you know, the currently projected Washington Command and Control Complex consists of the National Military Command Center (a soft installation in the Pentagon), the Alternate National Military Command Center at Ft. Ritchie, Md. (being hardened to withstand [redacted] of overpressure), the Emergency Airborne Command Post, the Emergency Command Post Afloat and [redacted] Studies indicate that the fixed facilities of this complex and their communications could be eliminated with reasonably high probability by a small number (6-10) of 10 megaton weapons, resulting in only the aircraft and the ship surviving. The aircraft, operating on ground alert at Andrews, would require 10 to 15 minutes to become airborne and another 10 minutes to fly beyond the lethal range of a 50 MT weapon if airburst over Andrews. The ship is 30 to 60 minutes flying time from Washington. Both times are in excess of the upper limit of expected tactical warning. Projected improvement in enemy weapons size and delivery means (sub-launched missiles) will further shorten this time. These considerations create serious doubt that currently projected facilities are keyed to today's threat, much less the threat of the 1970's, or that they adequately provide for protection of top civilian and military leaders who would be required to make and disseminate high level decisions in an emergency.

Studies of deep underground structures and analysis of weapons test data indicate that it is feasible to design and construct a command facility at depths of about 3,500 ft. so that it will withstand multiple direct hits of 200 to 300 MT weapons bursting at the surface or 100 MT weapons penetrating to depths of 70-100 feet. Extrapolation of weapons technology predicts that such weapons represent the upper practical limit to be credited to the enemy in the mid-1970's.

REPRODUCTION OF THIS DOCUMENT IN WHOLE OR IN PART IS PROHIBITED EXCEPT WITH PERMISSION OF THE ISSUING OFFICE

Page 1 of 7 Pages
Copy 22 of 25 Copies
Series "A"
CCS X-7300

EXCLUDED FROM AUTOMATIC REGRADING: DOD DIR 5200.10 DOES NOT APPLY

~~TOP SECRET~~

Illustration 3-3a: Memorandum for the President on the proposed Deep Underground Command Center. Notice the great depth under discussion here of 3,500 feet. *Source*: Lyndon Baines Johnson Presidential Library, Austin, Texas.

~~TOP SECRET~~

A Deep Underground Command facility would have two very fundamental functions: To protect key people with sufficient staff and data to render critical decisions, and to insure a survival of facilities to allow dissemination of these decisions. From a practical standpoint, the DUCC would serve these purposes only if the President and other high officials could move to it within the minimum warning time and if the movement could be made unobtrusively so that political and sociological factors would not make it undesirable. The very existence of a DUCC would also contribute in a very major way to the broad objective of deterrence of enemy attack by making a survivable control posture credible and by creating the impression of a strong will to fight if necessary.

The situations that might exist in the event of future war highlight the inadequacies of currently projected command facilities. If attacked we will be faced with an initial and critical decision in the selection of an appropriate response. The nature of the attack may not be clear at the outset, particularly if the attack is quite limited. Red China and even other lesser powers could have nuclear weapons in varying quantities by 1970. A single weapon on Washington could be a third party attempt to trigger general war between the two major powers by capitalizing on a crisis situation. It could be an accident. The appropriateness of our response here would be extremely critical. A prolonged continuation of mounting crisis between the U.S. and Russia might create situations in which we must consider initiating use of nuclear weapons. Our own confidence that we do possess the ability to control our forces would be a major factor bearing on such a decision.

If a major attack on the U.S. has been mounted there will be a sequence of key decisions vital to limiting the conflict and insuring as favorable an outcome as possible. Such important decisions would include utilization of withheld or residual forces, the direction of the course of military operations, and the timing of political negotiations when cessation is possible. Finally, steps for reconstruction of the country must be directed. Decisions of this import should be made by the President, or by responsible civilian officials succeeding him or designated by him, supported by competent senior military and political advisors. Accurate and continuous information as to the exact nature of the attack, the changing status and capabilities of surviving U.S. and allied military forces, factual assessment of the damage our forces have inflicted on the enemy and the damage we have suffered are the minimum essential elements of information required for such decisions. A DUCC would provide a logical, survivable node in the control structure at which the decision maker and his support could meet.

2

Illustration 3-3b: Memorandum for the President on the proposed Deep Underground Command Center. The interesting thing here is the discussion of surviving in an environment in which multiple nuclear strikes of potentially 200 or 300 hundred megatons would have occurred! The countryside above would have been a radioactive, lunar landscape for mile after mile with millions upon millions dead, and these planners were contemplating their own survival. Note that this facility would be for elite political and military figures only. Everyone else would face the incoming nuclear strikes on their own. *Source*: Lyndon Baines Johnson Presidential Library, Austin, Texas.

TOP SECRET

The primary alternative for the DUCC concerns size. Among the alternatives that have been investigated are austere and moderate size DUCC's at a depth of 3,500 feet. Cost estimates for these are as follows:

	Operating Space (Sq. Ft.)	Total Space (Sq.Ft.)	Cost a/ (Millions)	Lead-time (Months)	Number of People
Austere	5,000	10,000	110	47	50
Moderate Size	50,000	100,000	310	66	300

In these examples, multiple, dispersed and hardened communications exits are included. This arrangement would require the enemy to expend on the order of ten 100 megaton or tens of 10 megaton weapons for a high probability of destroying the facility or its communications. A pacing item in the construction of either size facility is the construction of a shaft to the facility depth. Almost two years is required to arrive at the 3,500 foot depth before construction of the main facility can begin.

The operational capsule of the DUCC would be located close to the Pentagon with an access elevator directly into the Pentagon building. Elevator shafts would also be built below the White House and State Department buildings to the facility depth with horizontal tunnels and rapid transportation to the facility itself. Access to these elevators would be gained from within these buildings allowing key individuals to proceed from their offices undetected to protected tunnel depths in less than ten minutes and to be in the facility in less than fifteen minutes.

Pros and Cons of Building A DUCC

The views expressed by the JCS on a DUCC in essence are that the provision of a highly survivable command center for the National Command Authorities is desirable. They feel that a DUCC would have certain advantages over existing alternate centers, particularly ready accessibility within anticipated warning time of a ballistic missile attack and feasibility of relocating key individuals to the facility inconspicuously during periods of crisis. The JCS is concerned over the size of the facility selected and feel that the austere DUCC is too small and should not be specified until functional studies and overall feasibility of the project as proposed have been determined. They also recommend that your views on the concept of a DUCC be obtained.

a/ 25 per cent special contingency has been added over normal contingencies.

3

TOP SECRET

Illustration 3-3c: Memorandum for the President on the proposed Deep Underground Command Center. Notice that elevator shafts would descend 3,500 feet underground from the White House, State Department and Pentagon, giving access to the 3,500 foot level in just ten minutes. Once at depth, personnel would travel through connecting tunnels to the DUCC. *Source*: Lyndon Baines Johnson Presidential Library, Austin, Texas.

~~TOP SECRET~~

There is little argument that construction of a DUCC is technically feasible and that it offers a unique capability in terms of accessibility and endurance to the President and key military and civil advisors. If constructed as proposed, so as to be accessible from within the White House, the Pentagon, and the State Department, it would provide protection based on 10 minutes or less of warning. It would have the potential of reducing significantly the problem of transition from peace to war. It offers the potential, under conditions of nuclear war, for establishment of a unified strategic command and control center under duly constituted political authorities. The nature of the DUCC, regardless of the initial size selected, lends itself to later expansion. Facilities at such depths are extremely costly, however, and require extensive construction leadtime. The need for and utility of the DUCC therefore merit close scrutiny. Among the primary arguments raised against the DUCC are:

1. The threat may outpace the survivability of the DUCC. The enemy may have the capability to develop weapons in excess of 300 MT yield or could elect to try to develop special penetrating weapons of greater than 100 MT yield.

a. Here there is, of course, much room for speculation. A primary question to be evaluated is would or will the enemy allocate the very sizable effort required to develop operational delivery systems and warheads specifically to destroy a DUCC? A countering consideration is that hard-point AICBM protection of the DUCC might just as likely outpace the threat, and with or without AICBM protection, the attacker would inevitably be in considerable uncertainty about his ability to destroy the DUCC.

2. Although the present National Military Command System will not protect the President, at least some general officers will survive. Other complementary alternatives exist such as prelocation of successors outside the Washington area or relocation of key people such as the Vice President to truly classified locations when strategic warning is received.

a. The fixed sites of the present system, as has been pointed out, are not very survivable. The mobile alternates are not accessible within expected tactical warning time. Consequently the personnel at the surviving command sites would probably not be the senior civil or military leaders. It is true that relocation of selected personnel does offer a potential enhancement of our survivability under any arrangement and should be pursued. We must recognize, however, that prelocation and relocation both require sufficient discipline to keep key people away from Washington during crisis situations. The very people we are trying to protect have the most need for direct access to the President. The feasibility of maintaining a truly classified location is open to question if survivable and secure communications are to be provided. Relocation

4

Illustration 3-3d: Memorandum for the President on the proposed Deep Underground Command Center. More discussion of survivability and weapons of apocalyptic destructive power. Also note the mention that the size of the facility could be expanded later. *Source*: Lyndon Baines Johnson Presidential Library, Austin, Texas.

to dispersed sites will be a lengthy and very visible procedure and it often may not be expedient to take this action. In the event of very serious situations, such as a limited war in which nuclear weapons have started to be used, the existence of a DUCC would permit relocation of a part of the key staff while holding the remainder in Washington.

3. If the DUCC were available, key people may not use it; or if an attack occurs during other than office hours, they will not be able to reach it.

a. Operational arrangements to provide for survival of key people are difficult at best. It would certainly be much easier to assign senior military and civilian personnel to emergency duty in the DUCC where they could continue to remain in close touch with critical activities than to assign them to remote relocation sites. The DUCC seems to offer the only feasible protection for key people in Washington when short tactical warning is received. Senior military or civilian key staff members who do gain access to the DUCC would make a likely contribution to the war effort especially considering that the only survivors at the Unified and Specified Command Headquarters are likely to be duty officers in mobile command posts.

4. If the enemy elects to attack Washington, he is irrevocably committed to full scale destruction and as long as a doctrine to insure U.S. retaliation is provided there is no real point in providing a survivable control mechanism at the national level.

a. In the event of a massive USSR attack which included directed destruction of Washington, there might be no reason for attempting to exercise a controlled response. However, as has been pointed out, a number of situations could arise in which Washington is attacked and control would still remain of paramount importance. Such situations would be: accidental attack on Washington during the course of a general war; third party attack; irrational or accidental small attack. Even in the event of directed major attack on Washington, the survival of key leaders is needed to terminate the war and direct reconstruction.

5. The political considerations of building a DUCC could be unfavorable. The impression that a major public expenditure is being considered for the survival of key government figures while little is being spent for protection of the public could have a significant impact on congressional review actions.

a. The introduction of this type of criticism of the DUCC program may be unavoidable. It is not feasible, and is in fact counter to the deterrent objective, to attempt to build so extensive a facility in secret. Instead, the command capability provided by the DUCC, its

5

Illustration 3-3e: Memorandum for the President on the proposed Deep Underground Command Center. Ponder the significance of point 5 above. The memo's author worries about the public's perception that Washington officialdom is concerned with saving itself with little regard for public safety and protection, in the event of a shooting, nuclear war. *Source:* Lyndon Baines Johnson Presidential Library, Austin, Texas.

functions of control and restraint and its deterrent value should be emphasized in mitigation against the view that protection of selected people is the motivating purpose. It should be noted that since the DUCC will not be in operation before about 1970, the incumbent administration will not "benefit" from such protection, and thus can consider the problem dispassionately.

Conclusions

The primary reason for proceeding with the DUCC would be a decision that there is a need for a survivable control capability to provide flexibility and latitude in dealing with contingencies, especially in escalating situations and a desire to convey an image of national will and determination during crisis and tension by making realistic provisions to fight if necessary. In short, the DUCC would contribute to a total impression of U.S. will and determination by making our command and control arrangements much more credible.

The DUCC offers unique capabilities for protection of key people and staffs. It provides accessibility compatible with tactical ballistic missile warning and convenience which could encourage inconspicuous relocation based on intelligence warning or developing crisis. Opinions admittedly vary as to the contribution of a DUCC to deterrence of massive attack and whether key people would in fact use the facility.

In considering the military needs together with the broader national issues I am convinced that a DUCC of at least the size of the austere proposal is required. I am satisfied that there are no unresolved technical problems that would prevent going ahead with construction of a DUCC at this time. There is general agreement that the austere and moderate sizes discussed earlier represent practical upper and lower size limits, but the exact size and configuration must wait for a more detailed functional analysis. A decision to build a DUCC now will save valuable time and will not preclude later size adjustment if future functional design definition so indicates. The two years required to dig the shaft permits deferral of final cavity design for a year or more without program slippage or wasted funds.

Recommendation

That a DUCC for the Washington area be approved now and that the austere size (10,000 sq.ft. - 50 man) DUCC be authorized according to the schedule outlined below with construction to be initiated beginning with FY 1965 funds. Design and engineering should be accomplished so that the facility could be expanded up to the size of the moderate DUCC (100,000 sq.ft. - 300 man), provided such a decision is made within a year following program approval.

6

Illustration 3-3f: Memorandum for the President on the proposed Deep Underground Command Center. Here you see that the memo's author, likely Secretary of Defense McNamara, has decided to give the green light to the project and begin construction of the 3,500 foot deep access shaft starting in FY 1965. I am guessing that something like the DUCC probably was built beneath Washington, DC and probably enlarged and/or deepened by the U.S. Navy in recent years. *Source*: Lyndon Baines Johnson Presidential Library, Austin, Texas.

TOP SECRET

	FY 65	FY 66	FY 67	FY 68	FY 69	Total
RDT&E	5M	2M	--	--	--	7M
Military Construction	23M	31M	35M	1M	--	90M
O&M	--	--	--	--	3M	3M
Procurement	--	7M	3M	--	--	10M
Total	28M	40M	38M	1M	3M	110M

The Secretary of State has seen this paper and concurs.

7

TOP SECRET

Illustration 3-3g: Memorandum for the President on the proposed Deep Underground Command Center. The projected cost was $100 million and construction completion was evidently expected in 1969. The memo states that the Secretary of State concurred with the recommendation. What did President Kennedy think about the plans? Did he agree? Did he object? He was killed two weeks later. *Source*: Lyndon Baines Johnson Presidential Library, Austin, Texas.

As a case in point, a memo from Walt Rostow to McGeorge Bundy, dated 16 January 1964 and carrying the subject of "DUCC Concept," states that Rostow believed the DUCC "study should go forward."

~~Personal-Confidential~~

DEPARTMENT OF STATE
Counselor and Chairman
Policy Planning Council
Washington

January 16, 1964

MEMORANDUM FOR MR. McGEORGE BUNDY
THE WHITE HOUSE

SUBJECT: DUCC Concept

Ken Hansen and Spurge Keeny were good enough to come over and brief me on the possible special study of the concept for the Deep Underground Command Center (DUCC).

My personal reaction is that this is a form of insurance for the continuity of civil government which prudent men should take, notably since the concept (and the Center itself) should relate to intervals of acute crisis, not merely to general nuclear war.

I do not know what the Secretary of State's reaction would be to a White House invitation to take part in such a proposed study; nor do I know whom he would assign to work with the White House and DOD.

My own recommendation, however, would be that the study should go forward and that the Department of State should participate. Moreover, I would be prepared to invest some serious S/P resources in the enterprise, if requested.

W. W. Rostow

DECLASSIFIED
E.O. 12958, Sec. 3.6
NLJ 97-84
By ics, NARA Date 6-18-97

~~Personal-Confidential~~

COPY LBJ LIBRARY

Illustration 3-4: Walt Rostow memo for McGeorge Bundy. The DUCC plan had high level support at the State Department. *Source*: Lyndon Baines Johnson Presidential Library, Austin, Texas.

This is significant, because Rostow was the Chairman of the Policy Planning Council at the State Department, and his views carried clout. Rostow had served President Kennedy as a senior adviser, and continued in a similar capacity in the Johnson administration.

Yet another high-level memo was written to McGeorge Bundy on 16 January 1964. This memo came from Carl Kaysen, who was Deputy Special Assistant for National Security Affairs to President Kennedy. I was not previously aware that Kaysen worked for Lyndon Johnson, and yet here is this memo for McGeorge Bundy, carrying a White House letterhead, written a couple of months after Kennedy's murder, and signed by Carl Kaysen. Clearly, Kaysen stayed on for a period of time after Kennedy died, perhaps specifically to work on the DUCC planning. I know the provenance of the memo – I received it directly from the National Archives and Records Administration (NARA) at the Lyndon Baines Johnson Presidential Library in Austin, Texas, so there is no question that the memo originated from Kaysen on 16 January 1964.

For our purposes, the memo is interesting because it shows that planning for the DUCC was going forward in the early months of the Johnson administration. Kaysen proposed that a committee be set up to deal with the DUCC, and suggested that McGeorge Bundy be Chairman. He also referred to having spoken with Harold Brown about the DUCC and the proposed DUCC committee. Harold Brown was Director of Defense Research and Engineering from 1961 to 1965, a period of time that overlapped both the Kennedy and Johnson administrations, so his inclusion in the DUCC policy discussion signals seriousness of intent on the part of those contemplating the construction of the DUCC. Later, during the Jimmy Carter years, Harold Brown was Secretary of Defense.

Defense

2

THE WHITE HOUSE

WASHINGTON

~~TOP SECRET~~ January 16, 1964

MEMORANDUM FOR MR. BUNDY

Harold Brown and I discussed the matter of the DUCC this morning. In view of the problems between the Secretary and the JCS, we agreed that the best way to handle the matter was to create a limited interdepartmental committee to study the problem from the point of view of the civilian top level of Government; and at the same time suggest to the Secretary of Defense that he request the Joint Chiefs to give their views on the nature of their relations with both the President-Secretary of State-Secretary of Defense level and the CINCs in a crisis situation toward the end of the sixties. The target date for this is the 15th of March.

The purpose of this would be to get the Chiefs to deal explicitly with their view of the relations between the top civilian level and the operational commanders during the period of crisis, and make clear both their ideas of what kinds of crisis situations they are thinking of and the amount and character of communication they would expect in both directions from and on location.

The interdepartmental study group would try to answer four questions, against the background of some likely scenarios of crisis in which a thermo-nuclear war is either imminent or has actually begun.

A. What would the utility of the DUCC be in this situation in the late sixties?

B. How big would the facility have to be in terms of the number of people it could hold to provide this utility?

C. Are there any unresolved technical problems which would have to be dealt with to make the installation effective?

~~TOP SECRET~~

Illustration 3-5a: Carl Kaysen memo for McGeorge Bundy. High level, White House planning for the DUCC continued early in the Johnson administration. *Source*: Lyndon Baines Johnson Presidential Library, Austin, Texas.

~~TOP SECRET~~ 2.

D. What would its relation be to the other elements of the National Military Command System (NMCS)?

Harold and I think the committee should be chaired formally by you, and that its members might be himself, Andy Goodpaster, Alex Johnson, Walt Rostow and Ray Cline. Spurgeon and I would join to represent you on the committee, and I could convene the meeting and act as Chairman in your absence. The main staff of the committee who would be available for full-time work would be furnished by Harold Brown's office. In addition, Jim Clark of BOB, who is knowledgeable on these problems, might serve on its staff.

E. Secretary McNamara might prefer to deal with this purely as an internal problem within the Department of Defense. However, the arguements for the other arrangement are convincing to Harold Brown and me. First, if there is to be a fight with the Congress, the President himself must be convinced of the need for the proposed facility, and this can best be done through the participation of his own staff. Second, there is not within the Pentagon the kind of experience that the White House-State-CIA are likely to have that is requisite to a thorough examination of the issues. While nobody has the relevant experience, the suggested group would come closer to having a basis forspeculation about it than any other we can think of.

CK.
Carl Kaysen

Illustration 3-5b: Carl Kaysen memo for McGeorge Bundy. High level, White House planning for the DUCC continued early in the Johnson administration. *Source*: Lyndon Baines Johnson Presidential Library, Austin, Texas

Finally, another high-level and TOP SECRET memo dealing explicitly with the DUCC was issued by the Bureau of the Budget (BOB). The memo is not signed, and while it carries no date, its content and context suggest that it also appeared in the late 1963-early 1964 time frame. The first thing that stands out about the BOB memo is that about 40% of it is whited out, presumably by the government censor who "sanitized" it when NARA declassified the document. If the DUCC was never built, why white out such extensive portions of the document? Especially in light of the fact that the seven-page memorandum for the President, dated 7 November 1963, which contains a fair amount of detail about the project, is not whited out at all? Was the DUCC project actually carried out, in one fashion or another? Is that why, more than 30 years later, an obscure document from the BOB was whited out by a government censor, prior to declassification? Or was the "sanitizer" just overzealous, and in a white-out mood?

TOP SECRET

Bureau of the Budget Staff Briefing Paper
On Secretary of Defense Draft Memorandum to the President
"National Deep Underground as a Key FY 1965 Budget Consideration"

I. Summary of Secretary's Memorandum

The 6-page memorandum on this subject should be read in full. It is similar to an earlier draft which you have seen.

The Secretary recommends that the President approve construction of an austere size [ ] Deep Underground Command Center (DUCC) to house [ ]

There would be access to the DUCC [ ]

[ ] The proposed 1965 funding to initiate the project would be $28 million ($23 million for construction of the main shaft and $5 million R&D). Total cost of the austere DUCC is estimated at $110 million spread mainly over the period 1965-1967. The facility would be operational in [ ]

An alternative presented is a moderate size DUCC with [ ] [ ] Total cost is estimated at $310 million, and it would be available about [ ]

1.5g

1.5g

II. Comment

The issue is the importance of providing for the survivability of the Presidency in a nuclear attack on the United States initiated with tactical (15 minute) warning. The President's Message to Congress in March 1961 stated:

"The basic decisions on our participation in any conflict and our response to any threat -- including all decisions relating to the use of nuclear weapons, or the escalation of a small war into a large one -- will be made by the regularly con-

TOP SECRET

SANITIZED
E.O. 12958, Sec. 3.6
NLJ 97-82
By Cb, NARA Date 12-17-97

Illustration 3-6a: Bureau of Budget memorandum on the DUCC. The date of anticipated completion of the DUCC project is whited out, but judging from the other memos cited above the year is likely to have been 1969 or thereabouts. *Source*: Lyndon Baines Johnson Presidential Library, Austin, Texas.

~~TOP SECRET~~

2

stituted civilian authorities .... The basic policies ... lay new emphasis on improved command and control -- more flexible, more selective, more deliberate, better protected and under ultimate civilian authority at all times."

Defense is investing large amounts of funds to achieve a survivable and controllable strategic response capability which under Administration policy would be subject to civilian control. It now appears that [redacted]

1.5g

The memorandum states on page 1 that there is "serious doubt that currently projected facilities are keyed to today's threat, much less the threat of the 1970's, or that they adequately provide for protection of top civilian and military leaders who would be required to make and disseminate high-level decisions in an emergency."

All other alternatives [redacted] appear less desirable than construction of a DUCC.

1.5g

On this basis, Division staff support the conclusion [redacted] Dr. Wiesner's staff participated in the early stages of the drafting of the memorandum and agree with the conclusions. State also concurs. [redacted]

1.5g

1.5g

III. Items for discussion

A. If a DUCC is to be constructed, what size should it be? [redacted]

1.5g

~~TOP SECRET~~

Illustration 3-6b: Bureau of Budget memorandum on the DUCC. Here the government censor got more zealous. This is interesting, seeing as more than 30 years had passed by the time the document was declassified. What does it matter if anyone knows what the memo said in 1964? Unless the facility really was built... *Source*: Lyndon Baines Johnson Presidential Library, Austin, Texas.

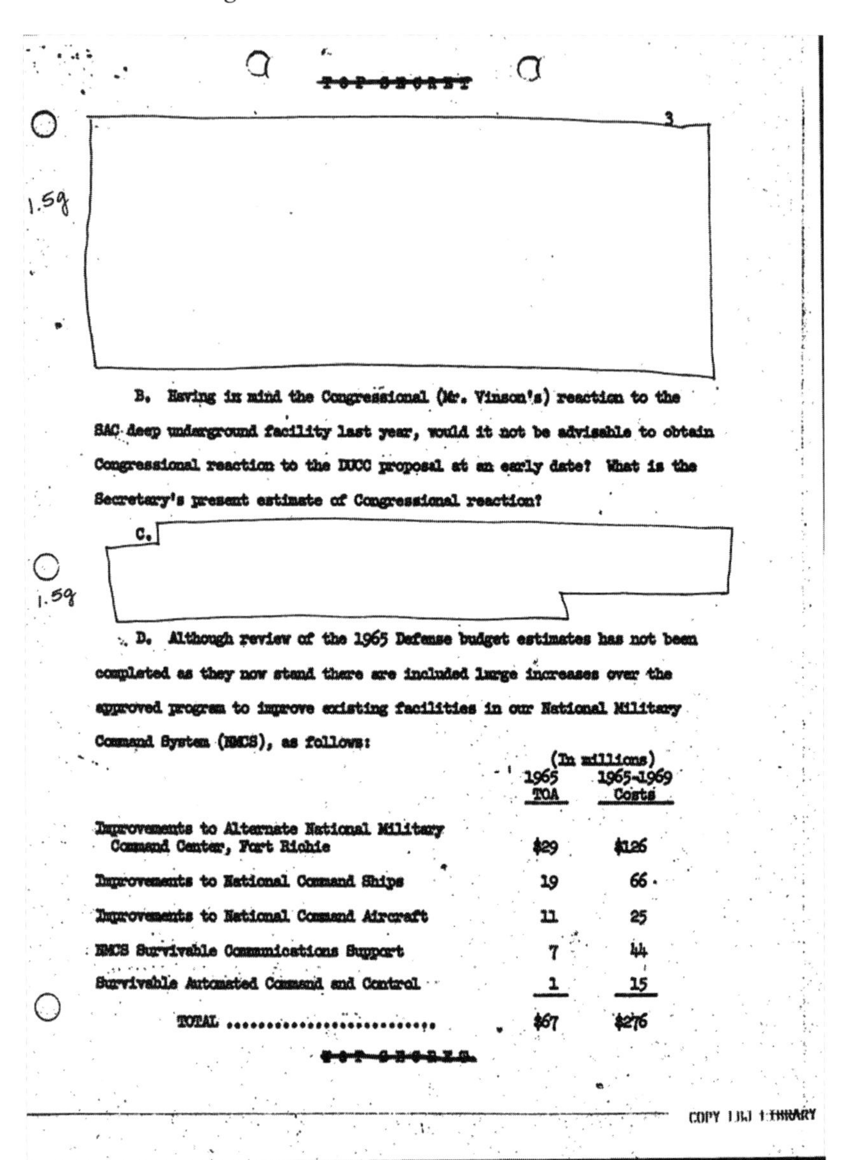

TOP SECRET

3

1.5g

B. Having in mind the Congressional (Mr. Vinson's) reaction to the SAC deep underground facility last year, would it not be advisable to obtain Congressional reaction to the DUCC proposal at an early date? What is the Secretary's present estimate of Congressional reaction?

C.

1.5g

D. Although review of the 1965 Defense budget estimates has not been completed as they now stand there are included large increases over the approved program to improve existing facilities in our National Military Command System (NMCS), as follows:

	(In millions) 1965 TOA	1965-1969 Costs
Improvements to Alternate National Military Command Center, Fort Richie	$29	$126
Improvements to National Command Ships	19	66
Improvements to National Command Aircraft	11	25
NMCS Survivable Communications Support	7	44
Survivable Automated Command and Control	1	15
TOTAL	$67	$276

TOP SECRET

Illustration 3-6c: Bureau of Budget memorandum on the DUCC. Here the memo plainly cites dollar figures for improvements to Fort Ritchie, command base for a major underground base for the Pentagon up on the Pennsylvania-Maryland state line, so why all the huge secrecy about dollar figures for the construction of the DUCC? *Source*: Lyndon Baines Johnson Presidential Library, Austin, Texas.

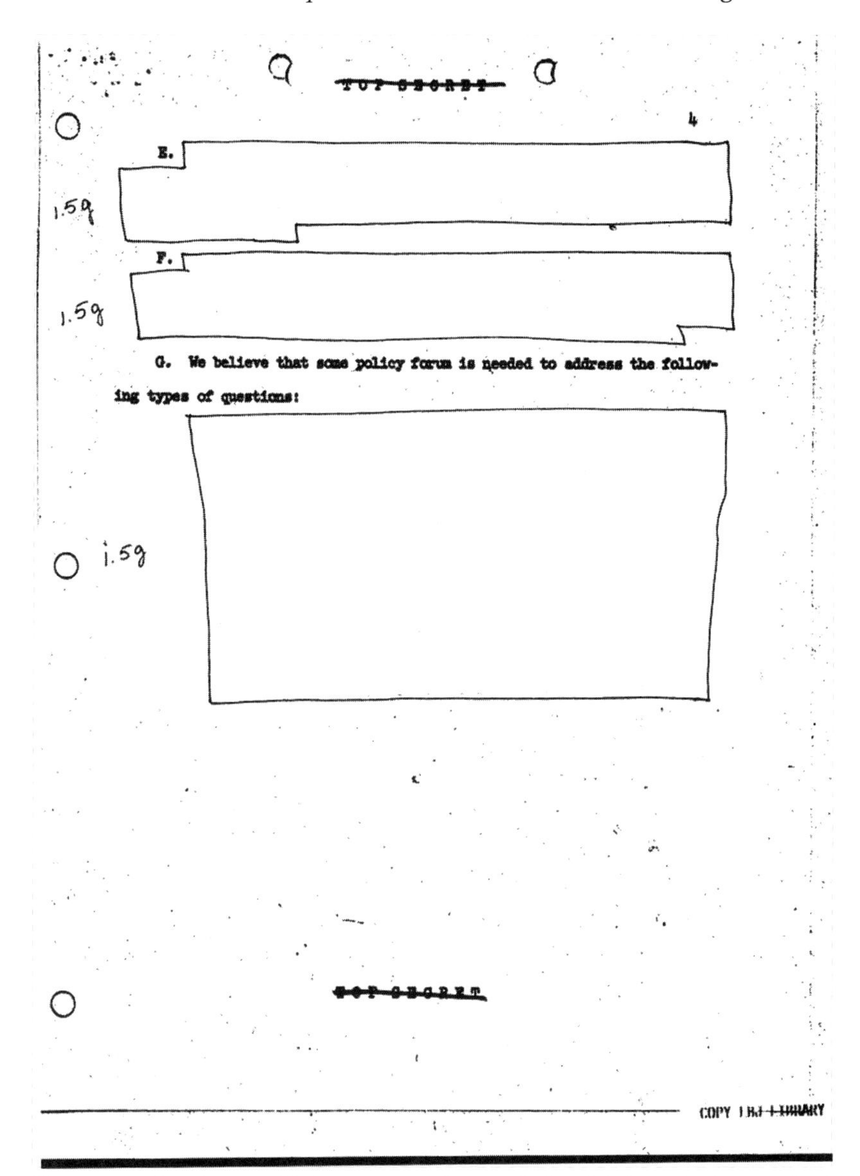
TOP SECRET

4

E.

1.5g

F.

1.5g

G. We believe that some policy forum is needed to address the following types of questions:

1.5g

TOP SECRET

COPY LBJ LIBRARY

Illustration 3-6d: : Bureau of Budget memorandum on the DUCC. Virtually the entire final page is whited out. The last page normally presents the conclusions of a document, so if nothing happened then it doesn't matter who knows what didn't happen, right? The obvious conclusion is that the plans for the DUCC went forward and there is a big labyrinth thousands of feet below Washington, DC. *Source*: Lyndon Baines Johnson Presidential Library, Austin, Texas.

One passage stands out. It reads as follows:

> All other alternatives [whited out] appear less desirable than construction of a DUCC.
>
> On this basis, Division staff support the conclusion [whited out] Dr. Wiesner's staff participated in the early stages of the drafting of the memorandum and agree with the conclusions. State also concurs. [whited out]

The tone of the extract suggests that the authors of the BOB memo agree with the DUCC plans. The last part of the memo carries the heading, *Items for discussion*, and is very heavily whited out. Evidently, even with the passage of the decades, the government censor felt that the material in the BOB memo is too sensitive to divulge publicly. And that seems peculiar, unless something like the DUCC really has been built.

Conclusion

It is my informed belief that something like the DUCC *was* built. I suspect that after the Kennedy assassination, the decision was made at a very high level of the government to go ahead and build a very deeply buried complex of tunnels and high-speed, ultra-deep elevator shafts way down below Washington, D.C.. That would be consistent with the content and tone of the declassified, TOP SECRET memos in the National Security archives at the Lyndon Baines Johnson Presidential Library.

It is interesting that the Washington, D.C. metro system began construction in 1969, not long after the date of the TOP SECRET memos discussed in this chapter. The Washington, D.C. metro is a fine transportation system, now carrying far more traffic than it was ever designed for. I have ridden the trains through its tunnels many times. But might its construction, which spanned some thirty years, have served, at least partially, as cover for other subterranean projects? Projects much more secretive and much more deeply buried than the metro tunnels and stations? I am remembering what my mysterious e-mail informant told me about secret tunnels under Washington, D.C. that have a secret connection to the metro train system, and the intriguing information that John Elvin reveals in *Hogs In The Tunnel!* Is there really a high-speed, pneumatic, tube train that rushes through a secret tunnel from the Pentagon to Dulles Airport, on the outskirts of the Washington, D.C. metro area? John Elvin says that there is, and though I can neither prove nor disprove his information, I am listening attentively to what he says. I am mindful, as well, of the findings of my Japanese counterpart, Shun Akiba, whose research suggests that there is a veritable labyrinth of hundreds of kilometers of secret tunnels beneath Tokyo.

Why should Washington, D.C. be any different than Tokyo? In fact, my research on Washington, D.C., like that of Shun Akiba on Tokyo, results in the conclusion that there is certainly a multi-level maze, if you will, beneath Washington, D.C.. The top level would be endless miles of utility tunnels for steam pipes, electrical, Internet and telephone conduits and cables, and sewers. Near the surface there would also be the various secret passages and tunnels used by politicians and others who want to keep a low public profile, for political, espionage or personal reasons. Some of these will be a couple of centuries old, others might be much newer, like the tunnel that the FBI and NSA constructed beneath the Russian Embassy. Included in this category might also be the subway system that runs between the U.S. Capitol Building and the office buildings used by U.S. Senators and Congressional Representatives.[26] Then there are abandoned train and trolley tunnels from years long ago, here and there across town, now closed off to public access. A little bit deeper

---

[26] "The Unites States Capitol Subway System," http://www.clouse.org/capitol1.html, 2005.

run the tunnels of the modern metro system, whose trains carry thousands of passengers back and forth beneath the city every day.

And farther down below the metro tunnels? What lies down there? Well, the declassified 1960s-era documents from the Kennedy and Johnson administrations reveal official interest at a TOP SECRET level in constructing a huge project at the 3,500 foot depth, with high-speed, deep elevator and tunnel access from the Pentagon, White House and State Department. That's way down there, to be sure, and it may seem beyond the pale to the uninitiated in these matters. However, my research shows that modern technology is easily capable of construction at that depth.

And not to forget all of those TOP SECRET explosions and "classified" rumblings of diesel generators over at Dick Cheney's mansion on Massachusetts Avenue, involving some sort of U.S. Navy "utility and infrastructure" work. Most people do not use high explosives to remodel their home, but, hey, it looks like ex-Vice President Cheney did. Very curious.

What to speak of the "utility upgrade" that was done down on the banks of the Potomac, again by the U.S. Navy, and again most "hush-hush," though not too "hush-hush" for the *Washington Post* to reveal that the project entailed 300 days of secret digging and that one of the companies involved was Kiewit, a company with known tunnel boring and underground construction expertise. Ever more curious.

Add it all up, and the odds are high for all manner of secret tunnels beneath Washington, D.C., starting just below street level and extending down, level after level, perhaps for thousands of feet, with ongoing expansion of the system, right down to the present day.

Chapter 4:

Subterranean Maglev Maze?

When I first began to research the issue of secret underground bases, I immediately started to encounter stories of secret underground tunnels of every description. At first I did not know what to make of the stories. I thought perhaps these sorts of tales were urban myths, tall tales spun up out of thin air by unsophisticated, untutored people. But as time went on I heard more and more of them. When I began to write and speak publicly about my research I ran into even more of these accounts. Some of what I heard was alleged to be first hand, some of it second hand and some of it was just out there in the cultural stew of late-20th Century- early 21st Century America.

I received a hand written letter from a long distance trucker telling me of a delivery he had made in Texas, for which he was required to drive some seventy miles through a maze of underground tunnels to deliver his load to a subterranean loading dock. The story seemed improbable to me, but then I had other people who were telling me about an underground interstate highway system. One man who grew up in the Boulder, Colorado area told me that it was believed by his teenage social circle that there were secret tunnels running eastward from Colorado into Kansas. Another correspondent assured me that she and her husband had seen a convoy of trucks in the upper Midwest drive down into an underground entrance and disappear from view.

And what to make of the persistent rumors of a secret subterranean, high speed, magnetic levitation (maglev), vacuum tube, shuttle train network said to crisscross the North American continent deep underground, and maybe the entire planet, including under the oceans?

I can only say that in the early 1990s the whole concept of a secret system of tunnels underlying the landscape seemed incredible to me. I considered myself an informed person and presumed that if such things existed I would certainly know about the whole affair. But then I had no knowledge at the time of the stupendous sums of money available in the black budget. I had a very rudimentary knowledge of how special access, compartmentalized programs and projects work. I did not know how powerful and sophisticated underground excavation technology was. And I did not yet fully appreciate how massively corrupt and deeply criminal the American system had become.

Now I better understand what is possible. I have a much better grasp of the advanced technology, the massive black budget funding, and the extreme human compartmentalization in the world of special access, clandestine programs. I comprehend now how profoundly compromised the entire system is, and how the charade presented to us in school, on television, at church, and by the government is an illusory mix of feel-good, mind numbing propaganda to keep us fat, happy, and blissfully dumb.

The Genius of Alfred Ely Beach

Most people today have never heard of Alfred Ely Beach, but in the post-Civil War era of the 19th Century he was well known in American scientific circles. Alfred Beach was the editor of the

Scientific American, which was, both then and now, a prominent American scientific journal. Even less well known is that Alfred Ely Beach was the constructor of the first subway line in New York City, and that he did it all on the sly, effectively operating one of the first clandestine, compartmentalized, "black" projects in the United States.

In 1869 Alfred Beach privately hired a team of excavators and tunnelers who very secretly carved out New York City's first subway line beneath Broadway. The way in which he carried out the operation was very clever. The work was done at night, under cover of darkness, going out from the basement of Devlin's Clothing store. Beach independently invented a tunneling "shield" that enabled the tunnel to be built at a rapid rate. When the project was finally revealed to the public it was a marvel of engineering and staid elegance. Beach spared no expense.

The subway line was pneumatic. The passenger car was propelled through the tube by a large fan, which was itself powered by a steam engine. Beach offered rides to the public for several weeks and while his invention was wildly popular, the corrupt nature of New York politics prevented his breakthrough from being adopted and extended to the whole city as a mass transit mode. Decades were to pass before work was begun in earnest on the extensive New York City subway system we know today.[1]

The previous year, Alfred Beach had published a book, *The Pneumatic Dispatch*,[2] on the technology of pneumatic tube transport. He was much taken with the great advancement of this technology in Great Britain. In 1864 a ¼ mile underground passenger line was

---

[1] Waldemar Kaempffert, "New York's First Subway," *Scientific American* 106, no. 8 (24 February 1912): 176-177.

[2] Alfred Ely Beach, *The Pneumatic Dispatch* (New York: The American News Company, 1868).

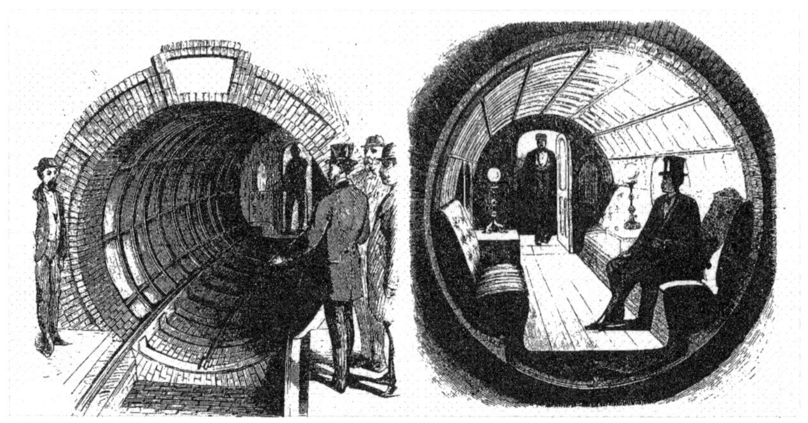

Illustration 4-1: Alfred Ely Beach's 1869 pneumatic subway tunnel under Broadway, in New York City. *Source: Scientific American,* 1912.

inaugurated from Sydenham to the armory near Penge Gate. Other pneumatic tube systems were in use, or under construction in Great Britain, for delivering mail and/or carrying human passengers. In 1868, a pneumatic system was under construction beneath London's Thames River to carry passengers underground, from one side of the Thames to the other. In the same period, the postal service was using pneumatic transport to move the mail in London.

Remember that Beach was the editor of the *Scientific American* and was perfectly up to date on the latest advances in science and technology. He was far ahead of his time in grasping the potential of this new technological development. Beach wrote:

> The Pneumatic Dispatch presents theoretical facilities for obtaining high velocities. Air rushes into a vacuum at the rate of 700 miles an hour. If the proper mountings for a light piston and finish for the tube could be realized, it is supposed that a velocity of piston approaching this might be obtained.[3]

Understand that Beach wrote these words in 1868. In his conception, the passenger car is the "piston," because it moves

---

[3] Ibid.

through the vacuum of the tunnel, as a piston does through the cylinder of an engine. He was saying that if he could make the passenger car light enough and make the tunnel smooth enough, that he could make a really fast, underground tube train, traveling at speeds approaching those obtained by commercial airliners today – and he was saying this in the years right after the American Civil War. Referring to one of the British pneumatic train systems, he went on to say that he thought that with improved efficiencies,

> ... a speed of 100 miles an hour may be safely realized, or four times the average speed of many of our best railroads.[4]

And Beach elaborated:

> In the pneumatic system no locomotives are employed. Hence, the roadway and cars may be very light. The whole pneumatic way being under cover, the road-bed is preserved from damage by the elements, and the transit of the cars is not impeded by snow, ice, floods, or falling rocks.[5]

In other words, in the years immediately following the American Civil War, Alfred Ely Beach had already grasped the advantage of very high speed vacuum tube trains, traveling at hundreds of miles per hour through underground tunnels. And he even managed to secretly build a simple, functional, working prototype for public use in mid-town New York. He understood what was technically possible – even with 19th Century steam-powered technology.

Later Developments

Although the use of pneumatic tube transport for human passengers was discontinued with the passage of the years, as the 19th Century progressed postal services made great use of pneumatic tube technology to move and deliver the mail. Large American cities such

---

[4] Ibid.

[5] Ibid.

as New York, Philadelphia, Boston, Chicago, and St. Louis installed underground tube systems that moved millions of letters. New York City only stopped moving mail around the city with pneumatic tubes in 1955. And in Europe, the Siemens Company of Germany had by 1865 installed pneumatic mail tube systems in Vienna, Paris and Berlin. The 250 mile-long pneumatic mail tube system in Paris, France was moving mail well into the 1970s.[6]

In the 20th Century, in the former Soviet Union, engineers employed pneumatic tube technology to move raw materials in the mining and quarrying industries. Their calculations are said to have "shown that pneumatic tube containerized transport systems are very efficient." An experimental system in the Republic of Georgia carried 25 ton loads of gravel through a pneumatic tube at speeds up to 45 kilometers per hour.[7]

In the United States today, pneumatic tube systems are still in use in the banking industry to move documents and money. Though it used to be more common in the 20th Century, there are still large, retail stores and office buildings that use pneumatic tube systems to move documents from one part of a building to another, from office to office, from clerk to clerk. The technology is proven. It works. It has been around for a couple of centuries now. As I shall show, these facts are surely well known in governmental, industrial and scientific circles, and almost certainly in military research circles, as well.

The Astonishing Rohrbahn of Hermann Kemper

During the Third Reich, a German engineer named Hermann Kemper patented a revolutionary new technology that he called the *Rohrbahn*.[8] The Rohrbahn was a magnetic-levitation train designed

---

[6] C.H. Vivian, "Early Pneumatic Tubes," *Compressed Air* 77, no. 1 (January 1972): 8-9.

[7] "Airborne in a Tube," *Compressed Air* 79, no. 8 (August 1974): 18-19.

[8] Die Rohrbahn, Reichspatent Nr. 643316, 1934.

to zoom through underground vacuum tunnels at hundreds, or even thousands of miles per hour.[9] The idea was that by suspending the passenger cars on a magnetic field and propelling them through tunnels from which the air had been evacuated, both the friction of wheels and rails (as in the case of conventional trains), and that of air itself could be negated. As a result, much higher efficiencies and speeds could be realized.

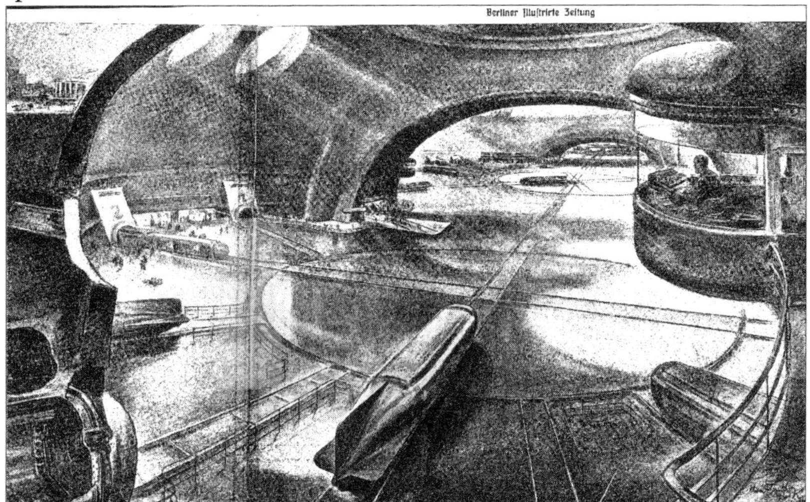

Illustration 4-2: Artist's conception of Hermann Kemper's *Rohrbahn. Credit: Berliner Illustrirte Zeitung*, 1938.

Kemper began his research in 1934 and continued right through WW-II and into the post-war period. In 1938, he envisioned a high-speed, underground, maglev tube train system radiating out from Berlin to many other European cities, carrying passengers at 600 mph (1000 km/h). However, Kemper was even considering speeds far higher: in the range of 1000 to 3000 km/h for the Rohrbahn.

It is not known how far the German engineers of the Third Reich carried Kemper's ideas. Certainly, Nazi engineers built

[9] "Reichspatent Nr. 643316 ! Die 'Rohrbahn'," *Berliner Illustrirte Zeitung* 47. Jahrgang, Nr. 33 (18 August 1938): 1246-1247.

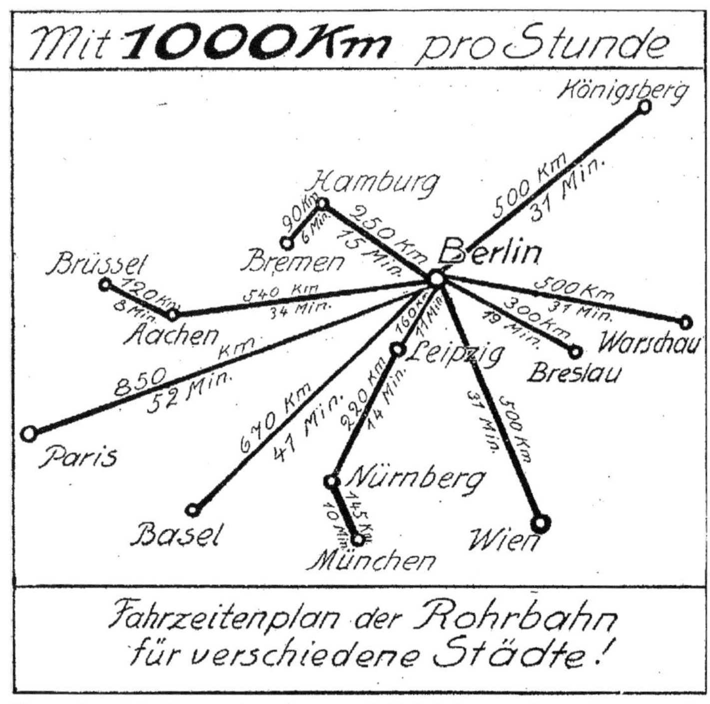

Illustration 4-3: Proposed underground Rohrbahn maglev train system in Europe. *Credit: Berliner Illustrirte Zeitung*, 1938.

impressive underground workings during WW-II, including at least one tunnel network that I am aware of, in what is now western Polish territory. In the so-called Miedzyrzecki Fortified Region, the Nazis built an extensive system of underground fortifications that they called the Regenwurmlager. The complex included tunnel systems up to 50 kilometers long, connected by electric trains running through subways buried 30 to 50 meters below the surface.[10]

It is also not clear whether Project Paperclip considered Kemper's work to have any merit, whether he was debriefed by Allied personnel, or whether Allied powers obtained any of the technology he had developed. Certainly, myriad other scientists and engineers (and their

[10] Paul Stonehill, "Secrets of the Regenwurmlager," *FATE* 55, no. 10 (November 2002): 28-33.

research) were scooped up by Project Paperclip after WW-II and utilized by American interests. However, Kemper's work did not go unnoticed in Europe, and it is my educated observation that it was taken very seriously. I strongly suspect that something like the underground, high-speed, maglev network that he proposed during the 1930s is now in use, both in Europe and in the USA, and perhaps elsewhere, too.

Illustration 4-4: Rohrbahn maglev passenger shuttle. Note that the car is suspended above the tunnel floor on a magnetic field. There is a partial vacuum in the tube. *Source: Berliner Illustrirte Zeitung*, 1938.

I have a fact sheet on Hermann Kemper and his Rohrbahn that I obtained from MVP GmbH, which is a subsidiary of the Deutsche Bahn AG and Lufthansa Commercial Holding GmbH.[11] This clearly indicates awareness of the importance of Kemper's work by the German transportation industry. In fact, Germany has been at the forefront of magnetic-levitation transportation R&D in the post-WW-II period.

Throughout the post-war years German companies and government agencies have continued to refine and advance high speed, maglev technology. Siemens, Thyssen, Krupp, Daimler Benz AG, the Deutsche Bahn and Lufthansa all have played leading roles in developing and furthering German maglev technology. In 1999 a firm named Transrapid International GmbH was formed by the

[11] Hermann Kemper: "Vater der Magnetbahn," undated factsheet from MVP Besucherzentrum Lathen and Transrapid Versuchsanlage Emsland.

ADtranz, Siemens and Thyssen companies. Transrapid International was created as a corporate entity to bring to fruition the Transrapid Maglev System, about which I have more to say below. During the last decade, Transrapid maglev agreements were signed by German government transportation officials with transportation agencies in China and the United States. In 2004 Siemens, ThyssenKrupp and Transrapid International put the finishing touches on a high speed maglev shuttle train in Shanghai, China that is now in use as public transit. Similar plans are on the drawing board for projects in both the USA and Germany.[12] The Siemens and ThyssenKrupp companies are co-founders of Transrapid International and are actively involved in planning high speed maglev train routes between German cities.[13]

Swiss Metro

Interestingly, an underground system of partial vacuum, maglev train tunnels – very closely resembling the system that Hermann Kemper proposed 70 years ago – is now in the R&D stage in Switzerland. The Swiss Metro Company has invested tens of millions of dollars of planning and research into a subterranean maglev train system linking all of the major cities of Switzerland, with cross-border connections to the transportation network in France, Italy and Germany. The plans call for the passenger trains to levitate magnetically and zoom through partial vacuum tunnels deep beneath the Alps at speeds of 300 mph.[14]

---

[12] "A History of Magnetic Levitation Transportation Technology," at http://www.transrapid-usa.com/content_usa_main.asp, 2004.

[13] "Transrapid International: A Joint Company of Siemens and ThyssenKrupp," at http://www.transrapid.de/en/index.html, 2004.

[14] See http://www.swissmetro.ch, 2009 and http://www.swissmetro.com/sito/default_eng.htm, 2002. Also http://www.geneva.ch/SMBern.htm, 2002.

Maglev Around the World

In recent years maglev planning in the USA has gotten a boost from the establishment of the Maglev 2000 company, which promotes development of maglev transportation worldwide and in the United States. Two of the principals, Gordon Danby and James Powell, patented superconducting maglev technology back in the 1960s.[15] In 2000 Lockheed Martin and Transrapid International-USA (a subsidiary of Transrapid International GmbH) signed an agreement to develop maglev technology in the USA. There were plans for maglev transit systems in seven different areas around the country.[16]

In Japan, maglev R&D is also well advanced. Japan National Railways has a 20 kilometer maglev test track in Yamanashi Prefecture, north of Mount Fuji. Test speeds as high as 581 km/h have been obtained for a manned, three-car train in 2003.[17] There are plans for a high speed, 300 mile maglev line between Osaka and Tokyo, 60% of which would be through deep underground tunnels.[18]

And then there is the proposed 4,000 mph, transatlantic maglev system of Ernst Frankel and Frank Davidson, retired MIT professors. In their system, a neutrally buoyant vacuum tube would be suspended about 150 to 300 feet beneath the ocean surface, securely

---

[15] Maglev 2000, http://maglev2000.com, 2009.

[16] "Lockheed Martin Teams with Transrapid International to Develop High-Speed Maglev Train projects in U.S.," http://www.lockheedmartin.com/news/press_releases/2000/LockheedMartinTeamsWithTransrapidIn.html, 2009.

[17] "Overview of Maglev R&D," http://www.rtri.or.jp/rd/maglev/html/english/maglev_frame_E.html

[18] "Maglev in Japan," http://www.maglev2000.com/today/today-03.html, 2002; http://www.rtri.or.jp/rd/maglev/html/english/maglev_ frame_E.html

tethered to the ocean floor. Transit time from New York City to Paris would be one hour.[19]

So Is There Really A Secret Underground High-Speed Maglev Tunnel System – Or What?

The rumors of a clandestine, underground maglev train system deep beneath the continental USA have been around now for a good 20 to 25 years. I first heard them about 20 years ago, in the late 1980s. Rumors have appeared on the far-flung corners of the Internet alleging that secret, underground train tunnels run from Mercury, Nevada to Groom Lake/Area 51, northwest of Las Vegas, and that a high speed tube shuttle runs from Coronado Naval Amphibious Base in California to Groom Lake.[20]

Such a system has not yet been confirmed to exist. However, it is definitely true that a huge amount of R&D work has been done on maglev tube train technology, beginning with Hermann Kemper's work in the Third Reich, all the way back in the 1930s. The R&D has gotten very sophisticated over the years and there are functioning, high-speed maglev trains in use today, such as the one in public use in Shanghai, China. Some of the world's most technically advanced corporations and transportation agencies are involved in maglev technology – Siemens, ThyssenKrupp, Lockheed Martin, Japan Railways, the Deutsche Bahn. With respect to the possibility of a secret, underground train system in the USA or elsewhere, I would remind you that there are literally trillions of dollars of public funds that are missing and unaccounted for in the American

---

[19] Carl Hoffman, "Trans-Atlantic Maglev: Vaccum Tube Train," http://www.popsci.com/scitech/article/2004-04/trans-atlantic-maglev, 2009; Von Walter Jäggi, "Mit der Rohrpost blitzschnell nach Amerika," http://www.tagesanzeiger.ch/dyn/leben/wissen/390805.html, 2004.

[20] http://www.ufomind.com/area51/list/1997/feb/a01-002.shtml, 2002 and http://www.ufomind.com/area51/list/1997/feb/a01-002.shtml, 2002.

system.[21] So funding for a secret project of any size whatsoever is not in question. And with the compartmentalization that is rampant in the military-industrial complex, I cannot categorically rule out the possibility that such a system could exist and be kept (mostly) secret.

My suspicion is that if the system was built, it took final shape in the post-Vietnam War era. That is the period when the maglev technology was starting to mature, and when tunnel boring technology became more sophisticated and capable of relatively rapid advances in deep, underground rock.

If such a system exists, it is highly likely that some (even many?) of the agencies and corporations mentioned in the pages of this book would have some role in its construction and operation. I reiterate that I have no solid information that such a system has been made, but hasten at the same time to tell you why I think that it just might exist.

Robert Goddard's Vacuum Tube Shuttle

Let me start by saying that, starting with Hermann Kemper's work in the 1930s, a slew of patents exist for maglev train systems, including those that zoom at extremely high rates of speed through vacuum tube tunnels at very great depth underground.

In 1944, Robert Goddard, the father of modern rocketry,[22] filed a patent application for a high speed, electromagnetic, vacuum tube transportation system. But in 1945 he abandoned that application and applied for a closely related patent entitled, *Vacuum Tube*

[21] "The War On Waste: Defense Department Cannot Account For 25% Of Funds — $2.3 Trillion." CBS News, Los Angeles, Jan. 29, 2002. http://www.cbsnews.com/stories/2002/01/29/eveningnews/main325985.shtml

[22] Interestingly, Robert Goddard did his most important work in Roswell, New Mexico, during the 1930s. Yes, that Roswell, New Mexico, just ten years before the now famous UFO crash-retrieval operation and subsequent cover-up by the U.S. Military. See: http://en.wikipedia.org/wiki/Robert_H._Goddard, 2009; and Jeffrey Kluger, "Robert Goddard," at http://www.time.com/time/time100/scientist/profile/goddard.html, 29 March 1999.

Transportation System. He described it as:

> ...a system of transportation in which a car containing goods or passengers is moved at a high speed through a transportation tube which is maintained under a substantial vacuum.[23]

In the text Goddard described propelling and supporting the car by means of "fluid pressure," said fluids specified to be gaseous in nature.

In another patent entitled, *Apparatus for Vacuum Tube Transportation,*[24] Goodard gave an idea of just how fast he envisioned such vacuum tube shuttle cars to travel. Let me cite his words directly:[25]

> A forward or horizontal acceleration of 4.89 g's would make the time of transit from Boston to New York (200 miles) 2 minutes 46 seconds, and from New York to San Francisco (2500 miles) 10 minutes, 4 seconds. This results in a total force of 5 g's acting on the passenger, which is assumed to be the maximum acceleration that can be borne with comfort and safety over a considerable period of travel.

Do the calculation. Goddard is describing a system designed to zoom coast to coast at hypersonic speeds, in a near vacuum, approaching 15,000 mph. That's almost orbital velocity, inside a vacuum tube tunnel. Robert Goddard died in 1945, before his patents were granted. But his historical stature as the founder of modern rocketry was assured due to his trail-blazing work with liquid-fueled rockets. And his ultra-high-speed, vacuum tube shuttle train patent did not go unnoticed by another prominent American "rocket scientist."

Michael Minovitch's Deep Underground Tube Shuttle

In the summer of 1961, Michael Minovitch, a young mathematics and physics graduate student working summers at the Jet

---

[23] U.S. Patent 2 511 979, patented 20 June 1950.

[24] U.S. Patent 2 488 287, patented 15 November 1949.

[25] Ibid.

Propulsion Laboratory in Pasadena, California, made a revolutionary breakthrough in the propulsion of deep space probes. Minovitch's pioneering work made use of planetary gravity attraction to slingshot interplanetary probes from one planet to another.[26] It proved to be a huge breakthrough and opened up the outermost reaches of the solar system to exploration by deep space probes.

But Michael Minovitch did not restrict his theories on gravity-assisted propulsion solely to outer space transport. In the 1970s, the young scientist, like Goddard before him, turned his scientific expertise toward patenting a high speed, vacuum tunnel train system. While his design is not intended to reach the astonishing speeds that Goddard envisioned, it nonetheless has an impressively deep and fast capability.

Minovitch called his system a *High Speed Transit System*.[27] The patent describes a high speed, deep underground, gravity powered, magnetic-levitation transit system that would zoom beneath the surface at speeds exceeding those attained by commercial jet liners. The abstract provides the following description:

> A high speed ground transportation system, is suspended in an underground vacuum tube by a frictionless magnetic suspension system and propelled by gravity. The tubes are suspended inside deep underground tunnels from anchor points near each adjacent station and follow smooth catenary curves similar to the main suspension cables of a suspension bridge. Gravity propulsion is accomplished by allowing the vehicle to coast down the descending arc of the tube, during which time it is accelerated by gravity, and decelerating by gravitational braking while coasting up the tube's ascending arc.
>
> Thus, the trip is accomplished by transforming the vehicle's gravitational potential energy at one station into kinetic energy and back into gravitational potential energy at the next station. Excess kinetic energy

[26] For a comprehensive account of Michael Minovitch's discovery, please see http://www.gravityassist.com, 2009.

[27] U.S. Patent 4 148 260, patented 10 April 1979.

United States Patent [19]
Minovitch

[11] **4,148,260**
[45] *** Apr. 10, 1979**

[54] HIGH SPEED TRANSIT SYSTEM

[76] Inventor: **Michael A. Minovitch,** 2832 St. George St. Apt. 6, Los Angeles, Calif. 90027

[*] Notice: The portion of the term of this patent subsequent to May 4, 1993, has been disclaimed.

[21] Appl. No.: **682,085**

[22] Filed: **Apr. 30, 1976**

Related U.S. Application Data

[63] Continuation-in-part of Ser. No. 466,609, May 3, 1974, Pat. No. 3,954,064, which is a continuation-in-part of Ser. No. 438,230, Jan. 31, 1974, Pat. No. 4,075,948.

[51] **Int. Cl.²** **B61B 13/10;** B61B 13/08
[52] **U.S. Cl.** .... **104/138 R;** 49/68; 104/148 MS; 104/148 LM; 105/365; 114/335; 138/107
[58] **Field of Search** .... 104/138 R, 148 MS, 148 LM; 105/150, 365; 310/67 R, 74, 113; 244/161, 137 P; 49/68, 482; 61/69 R, 83; 114/16.6; 138/107

[56] **References Cited**

U.S. PATENT DOCUMENTS

432,615	7/1890	Henning	104/138 R
891,416	6/1908	Fenyö	104/138 R X
1,772,459	8/1930	Grieshaber	61/69 R X
2,114,038	4/1938	Updegraff	61/83
2,488,287	11/1949	Goddard	104/138 R UX
2,589,453	3/1952	Storsand	310/74 X
2,942,816	6/1960	Dustie	244/137 P
3,404,638	10/1968	Edwards	104/138 R X
3,610,166	10/1971	Ellzey	105/150
3,656,436	4/1972	Edwards	104/138 R
3,683,216	8/1972	Post	310/67
3,738,281	6/1973	Waidelich	104/148 MS X
3,899,979	8/1975	Godsey, Jr.	104/148 MS
3,946,571	3/1976	Pate et al.	61/69 R
3,952,976	4/1976	Fletcher et al.	244/161 X
3,954,064	5/1976	Minovitch	104/138 R

FOREIGN PATENT DOCUMENTS

1035764 7/1966 United Kingdom .... 104/148 MS

Primary Examiner—Francis S. Husar
Assistant Examiner—Randolph A. Reese
Attorney, Agent, or Firm—Christie, Parker & Hale

[57] **ABSTRACT**

A high speed ground transportation system, is suspended in an underground vacuum tube by a frictionless magnetic suspension system and propelled by gravity. The tubes are suspended inside deep underground tunnels from anchor points near each adjacent station and follow smooth catenary curves similar to the main suspension cables of a suspension bridge. Gravity propulsion is accomplished by allowing the vehicle to coast down the descending arc of the tube, during which time it is accelerated by gravity, and decelerating by gravitational braking while coasting up the tube's ascending arc. Thus, the trip is accomplished by transforming the vehicle's gravitational potential energy at one station into kinetic energy and back into gravitational potential energy at the next station. Excess kinetic energy arising from coasting between stations having different elevations is supplied or absorbed by on-board linear motor/-generators that provide supplementary propulsion or regenerative braking. These linear motor/generators draw and return energy to an on-board flywheel kinetic energy storage system. Passenger and cargo transfer between the vehicle's interior and station is made without removing the vacuum environment of the vehicle, by providing air-locks through the tube walls at the station.

54 Claims, 25 Drawing Figures

Illustration 4-6: Michael Minovitch's proposed deep underground, maglev, High Speed Transit System. Notice the long looping tunnel that dips deep underground and then rises back up near the surface again.

arising from coasting between stations having different elevations is supplied or absorbed by on-board linear motor/generators that provide supplementary propulsion or regenerative braking. These linear motor/generators draw and return energy to an on-board flywheel kinetic energy storage system. Passenger and cargo transfer between the vehicle's interior and station is

> made without removing the vacuum environment of the vehicle, by providing air-locks through the tube walls at the station.[28]

In other words the train floats on a magnetic field and slides downhill, inside a tunnel from which the air has been evacuated. The magnetic field that it rides on and the airless tunnel reduce friction to negligible levels, making it possible for the train to reach a very high rate of speed with a minimal expenditure of energy.

You can see echoes of Minovitch's earlier work at JPL where he used planetary gravity fields to accelerate space craft to high rates of speed in outer space – whereas here he is using the gravity of the Earth to accelerate a vehicle to a high rate of speed in inner space, in subterranean tunnels buried deep beneath the Earth's surface.

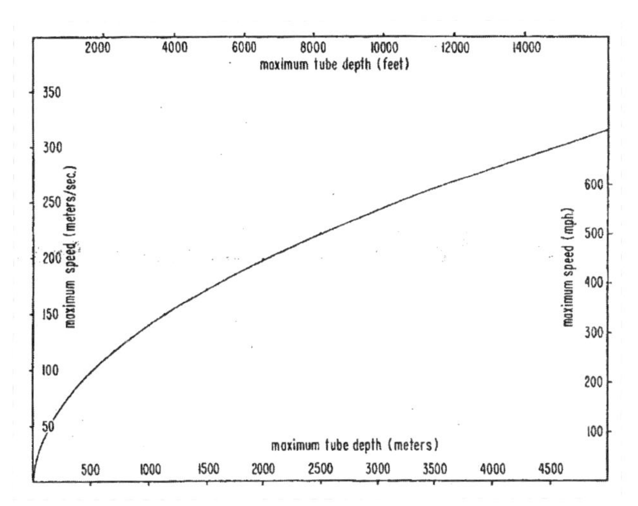

Illustration 4-7: Notice that in Michael Minovitch's deep underground, maglev, vacuum tube, train system the trains could reach speeds faster than those reached by jet airliners, at depths in excess of 14,000 feet. The deeper the tunnel, the faster the speeds. Using a back-of-the-envelope extrapolation, the data that Minovitch provide suggest that a speed of 700 mph can be attained with a maximum tunnel depth of 16,000 feet. *Source:* Michael Minovitch, U.S. Patent 4 148 260.

[28] Ibid.

The beauty of a system that uses this design is that once it is built it uses very little power. It is very efficient and practically runs itself. The patent makes clear that the operating depths could be very deep, indeed, approaching depths of three miles and speeds of 700mph.

I have no idea if something like this system has been built. There are certainly rumors floating around that a high-speed, deep underground, maglev train system is in clandestine use. These stories have been around for years and I have heard and read them from time to time over the years.

If you, the readers of this book, have firm evidence of such systems having been built and can tell us if these sorts of technologies are in operational use, I hope that you will come forward and reveal the truth to the rest of us. We, the general population, are paying for all of these clandestine, black operations. We deserve to know what is being done behind our backs, without our knowledge, with our money.

Robert Salter's Very High Speed Transit System

Robert Salter has also attracted a lot of attention with his own unique proposals for a Very High Speed Transit (VHST) system. In a 1972 study for the RAND Corporation, Salter set out a continental-scale, deep underground concept that his study succinctly described as:

> ...electromagnetically levitated and propelled cars in an evacuated tunnel.[29]

This study is especially noteworthy, in view of the role that the RAND Corporation has played over the last 60 years carrying out policy analysis and research at the highest levels for the American military, private industry, and now agencies and organizations in

---

[29] R.M. Salter, "The Very High Speed Transit System," RAND Corporation Report P-4874, August 1972.

other countries as well.[30] The Board of Trustees of the Rand Corporation are drawn from the upper strata of the American corporate, military, academic and high finance sectors.[31]

Salter stressed that his system would be very "conservative" of energy, in that the energy used to accelerate the train, riding on an electromagnetic field the way a surf boarder rides a cresting wave, would be stored in the train as kinetic energy, and then would be returned to the system as the train decelerated. When Salter described his system as a "Very High Speed Transit System" he was not kidding. He mentioned that:

> Speeds as high as 14,000 mph have been examined in studies by the Rand Corporation (in an example case of a direct link between Los Angeles and New York requiring 21 minutes transit time). The speeds required will certainly be on the order of thousands of miles per hour on the long-haul links.[32]

Shades of Hermann Kemper or Robert Goddard. Salter's projected system incorporated elements of the earlier work of both men.

Salter's system envisioned a network of VHST tubes crisscrossing the country. In his study, he strongly recommended placing the maglev system underground, where the trains would not be subject to interference with other environmental factors such as grade crossings, weather, or other infrastructure. He also advocated using the tunnel rights-of-way for other transportation, communications, industrial and utility systems.

Interestingly, Salter observed that it was not appreciably harder to tunnel at great depth underground than at shallower depths, and that it was therefore:

---

[30] RAND Corporation, http://www.rand.org/about/, 2009.

[31] RAND Corporation Board of Trustees, http://www.rand.org/about/organization/randtrustees.html, 2009.

[32] R.M. Salter, "The Very High Speed Transit System," RAND Corporation Report P-4874, August 1972.

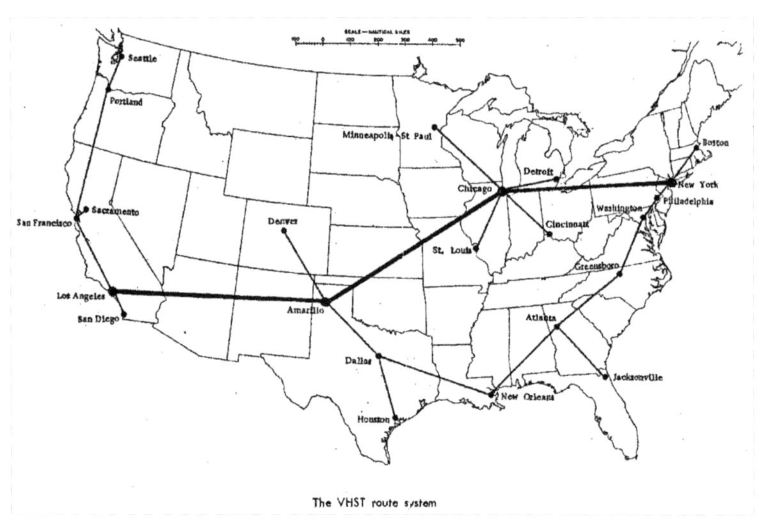

Illustration 4-8: Projected routes for VHST system. *Source:* R. M. Salter, RAND Corporation Report P-4874, 1972.

> ...possible to consider tunneling under deep water. A tunnel following the great circle route from Seattle to Paris would not require going under any ocean deeps, and in fact, would be under land masses most of the distance; the maximum depth requirement would be generally less than one mile.[33]

Of course, this is a topic also covered later in this book, notably in the parts dealing with the U.S. Navy Rock-Site report – which proposed siting permanent, manned bases at great depth beneath the sea floor, connected to sub-sea-floor tunnel systems potentially extending for hundreds of miles out to sea.

Have secret, sub-sea-floor, manned bases and tunnel systems been constructed? While I cannot provide conclusive evidence, I think that the answer is probably in the affirmative.

The High Speed Ground Transportation Initiative

As if all of the foregoing plans, studies, patents and information

---

[33] Salter, 1972.

were not sufficient to give pause to reflect on what may be secretly zooming beneath our feet, deep beneath the surface of the land and sea, there is the matter of the impressively extensive High Speed Ground Transportation Initiative. This began in the mid-1960s and continued through the 1970s, only to sputter out, and pretty much slip below the policy radar and out of public view in the 1980s.

But for a 10 to 15 year period, there was an intensive flurry of R&D in the USA, centered on the feasibility and desirability of high speed, underground, tube shuttle trains. The public literature that my research uncovered emphasizes the transportation problems of a growing nation, with increasingly congested urban corridors and the need to move passengers and freight expeditiously and efficiently around the country, from point to point.

The agencies, companies and individuals involved in the many articles, studies and reports that I have seen were a representative cross section of academic and technical experts, as well as corporate, engineering and military and government agencies involved in civil and military engineering and construction. TRW, Parsons, Brinckerhoff, Quade & Douglas, General Dynamics, Westinghouse, MIT, Southwest Research Institute, General Electric, The U.S. Commerce Department, The U.S. Transportation Department, NASA, The U.S. Army, The U.S. Navy, Allis-Chalmers, Battelle Memorial Institute, The Pennsylvania Railroad, Long Island Railroad – it was like a reunion of the military-industrial complex with the sole item on the agenda being the creation of high-speed, ground transportation in the USA. A great deal of the literature emphasized very high-speed, *underground*, tube train transport. Scores of plans, patents, reports, articles and studies were churned out – and then the hubbub all died away to a barely perceptible, background murmur.

Here's an example of the kind of thing I'm talking about. This is representative of the sort of information you will find when you dig into the technical literature on this topic.

TRW Systems prepared a study for the U.S. Department of Transportation in 1967 that dealt with several proposals for underground transit systems.[34] One was a novel concept put forward by aerospace engineering researcher and professor, Joseph Foa. Foa's idea was to send futuristic craft (Illustration 4-9) thundering through underground tunnels at "speeds equal to or greater than present day airliners..."[35] Foa proposed to propel his high-speed, tube train using either ram-jet, or magnetohydrodynamic technology, or both.[36] Is something like that operational today?

Another high-speed, tube transit system that TRW evaluated for the U.S. Department of Transportation was to operate in tunnels potentially dipping as deep as 3,000 feet underground using trains at speeds as high as 450 mph.[37]

In 1965, *Scientific American* featured an article by L.K. Edwards, in which he discussed a Northeast Corridor system that would have

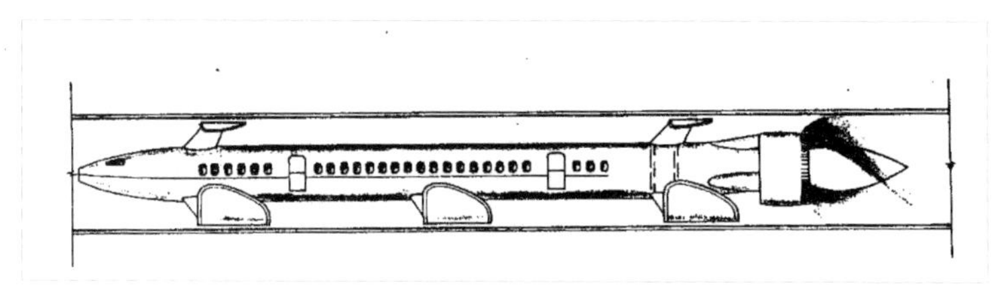

Illustration 4-9: Foa Tube System. *Source:* High Speed Ground Transportation System Engineering Studies Program, prepared by TRW for the Office of High Speed Ground Transportation, U.S. Department of Transportation, 1967.

[34] "Abstracts of Concepts of High Speed Ground Transportation Systems," no. 6730.01-107, High Speed Ground Transportation System Engineering Studies Program, Prepared by TRW Systems for the Office of High Speed Ground Transportation, U.S. Department of Transportation, 5 July 1967.

[35] U.S. Patent 3 213 802, patent granted 26 October 1965.

[36] Ibid.

[37] M. King and J.W. Smylie, "State-of-the-Art Tube Vehicle System," no. 06818-6042-R0-00, Prepared by TRW Systems for the Office of High Speed Ground Transportation, U.S. Department of Transportation, June 1970.

trains hurtling through underground tunnels between Washington, D.C. and Boston at cruising speeds up to 500 mph.[38] In Edwards' scheme, the trains would be propelled by one atmosphere air pressure through tubes from which the air had been evacuated – in essence a pneumatic system, very similar to what Alfred Ely Beach built under Broadway in 1869, but an order of magnitude faster. The basic mode of propulsion – differential air pressure -- would be the same as that which speeds a canister carrying your money and a bank deposit slip from your car window through a pneumatic tube to the drive-in-teller's window when you visit your local bank.

Edwards' system also incorporates some of the same, fundamental ideas that Michael Minovitch elucidates in his 1979 patent. To wit, he advocates employing the so-called "pendulum technique," wherein the tube train accelerates down a grade through a sloping, curved tunnel, under the influence of the pull of gravity, and plunges through the evacuated tunnel thousands of feet underground, and then as it comes back upwards again towards the surface, it gradually decelerates as the Earth's gravity slows it on the upward slope towards the destination terminal. Edwards saw his system dipping 3,500 feet deep and reaching speeds of more than 500 mph, using both gravity and one atmosphere air pressure for propulsion, making it extremely energy efficient and economical.[39]

And there's much more. In the mid-1960s, the International Society For Terrain-Vehicle Systems (ISTVS) was established with representation by engineering professors from several well-known universities, as well as representatives from the Office of Naval Research in Washington, D.C., from the U.S. Army Research Office in Durham, North Carolina, from Booz-Allen Applied Research in

---

[38] L.K. Edwards, "High Speed Tube Transportation," *Scientific American*, 213 no. 2 (August 1965): 30-40.

[39] Ibid.

Bethesda, Maryland, Grumman Aircraft, and Cornell Aeronautical Laboratory. The plain English translation is that this was a multi-agency working group established to coordinate research on high-speed, underground tube train systems. The Navy and the Army were represented, representatives from major university engineering departments were drawn in, Grumman Aircraft and the intelligence connection via Booz-Allen – long recognized as a front and planning organization for the federal "alphabet soup" agencies, including the CIA, NSA, etc. In 1967, the ISTVS established a journal dedicated in large part to publishing and promoting research and articles about high-speed, underground, tube train transit system technology and R&D, under the aegis of a High Speed Ground Transportation Committee. The initial officers of the ISVTS were Canadian, British and American.

The *High Speed Ground Transportation Journal* featured research on pneumatic tube transport carried out at Duke University and also had close ties with the U.S. Army Research Office in Durham, North Carolina. Other research for high-speed train systems was carried out at the U.S. Navy's China Lake Weapons Station's high-speed, rocket sled test track, testing electrical power pick-up at speeds over 300 mph.[40]

At the same time that ISVTS and High Speed Ground Transportation Journal were being established, the U.S. Department of Commerce sponsored a multi-agency panel dealing with many of the same issues, including the question of high-speed, underground, tube train systems.[41] The report clearly discussed designing and construct-

[40] Edward J. Ward, "Noise in Ground Transportation Systems," *High Speed Ground Transportation Journal*, 7 no. 3 (1973): 297-305.

[41] "Research and Development for High Speed ground Transportation," PB 173911, Report of the Panel on High Speed Ground Transportation, Convened by the Commerce Technical Advisory Board, U. S. Department of Commerce, Washington, D.C., March 1967.

ing a high-speed transportation system, much of which would be underground. In Appendix C the report's authors stated:

> Regardless of the system adopted it can be assumed that extensive tunneling will be involved. It is recommended that the Department cooperate closely with other agencies in planning and funding research in this area to provide an extra stimulus for rapid development.[42]

The antepenultimate paragraph in the report made the following, telling point:

> Portions or all of the HSGT system may need to be enclosed, probably requiring windowless vehicles.[43]

It stands to reason that if passengers are traveling at several hundreds, maybe even several thousand miles per hour, at depths of hundreds or thousands of feet underground, over distances ranging from hundreds to thousands of miles – and 99% or more of the trip is inside a sealed, vacuum tube 10 to 15 feet in diameter – then what's to see? Windows become superfluous. Remember that the Robert Goddard tube train was envisioned to traverse the continent in mere *minutes!* As was the Robert Salter tube train. Even with a couple of intermediate stops, Robert Salter's system would take only about 50 minutes to travel coast to coast. So why have windows? What could you possibly see, zooming along at thousands of miles per hour, thousands of feet underground, in a totally enclosed tube?

The Obvious Question

What it all comes down to, from the standpoint of this book, is whether there actually are secret underground tunnel systems with high-speed tube trains traveling through them, without greater public knowledge.

---

[42] Ibid.

[43] Ibid.

In my opinion, the answer is probably yes. My reasoning is based on several factors.

First, we now know that vast sums of money have been drained out of the public treasury for many years. The amount of money that is unaccounted for across the many federal agencies runs into the trillions of dollars. So the funding for a major, clandestine, civil engineering project is plentiful and available.

Second, we know that the military-industrial-espionage system is characterized by a pervasive, extreme degree of operational compartmentalization. The social technology for compartmentalizing people and concealing their organized activities is very smooth and sophisticated.

Third, civil engineering technology, and materials science have made great advances over the last half century. Tunnel boring machine technology is affordable and widely available. Many companies have extensive experience with mechanized tunneling and thousands of miles of tunnels have been bored. The technology is reliable and proven. Many alloys, composite materials, and plastics have been developed by industry that make high speed operation more feasible now than in past decades.

Fourth, impressive gains have been made in electronics and electrical engineering, making automated control of high-speed, high performance systems more reliable.

Fifth, a general dumbing down of the populace has made social control of the public easier in some respects than it was in past periods. Slick propaganda and public relations-style marketing of corporate and government disinformation have created passive acceptance of a corporate-government manufactured pseudo-reality that is believed with little question by a large percentage of the population. As a result, it becomes easier to disguise such a project, if indeed it has been built.

Sixth, a number of anecdotal accounts and rumors of just such a system have surfaced over the years, lending support to the supposition that it probably exists.

Seventh, as I have shown, there is an impressive body of patents, articles, reports, research, studies and federal documents that plainly discuss such systems. The open literature paper trail shows unambiguously that the technology for high-speed, underground, tube train systems has been under serious R&D for at least 75 years now.

For all these reasons, in my view the odds that something along those lines has been actually built and is in secret use must be greater than 50%. I think it's likely that the U.S. Government, the Pentagon and the major corporations have deceived the general public. Given the nature of the evidence I have adduced, I also admit of the possibility that the system could extend internationally, maybe even globally.

At the end of the day, how deep does the rabbit hole go?

Chapter 5:

Into the Abyss

It is important to understand that modern technology permits construction, maintenance and access to manned bases and tunnels at very great depth beneath both land and sea, at depths of thousands of feet and even miles.

People who are not up to date on advances in cutting-edge civil and marine engineering technology do not encounter in their daily life the extreme conditions under which major underground and deep sea projects are carried out. In this chapter, I will present a representative selection of information that is easily available in the open engineering, military, corporate and civil government literature. To the uninitiated, some of what follows might seem improbable or impossible, and that is why I present this material. We are living in a world where a very broad segment of the global population really has no idea of the sort of technology that has been developed, and of the policies and projects that are being carried out without their knowledge or consent on the very planet that they inhabit. Keep in mind that what follows comes from the open literature, from publicly available sources. I have no doubt that the technology in the world of compartmentalized, clandestine operations substantially exceeds the capabilities of what I describe – though the technology that is available in the public domain is already jaw dropping.

How Deep Beneath the Sea?

Because of improvements in diving suit technology, the U.S. Navy and private industry now have the capability to enable divers to work at 2,000 foot depths for hours, without having to undergo lengthy, days- or weeks-long pressurization or depressurization procedures. The so-called Hardsuit™ 2000 can deliver divers to depths of 2,000 feet within 20 minutes, and permit them to work almost 8 hours at a stretch. The suit makes deep water divers resemble the bulky-looking Michelin Man, of Michelin tire advertisement fame, but it represents a revolutionary advance in deep diving technology, permitting divers to quickly reach great depth and linger there for many hours at a time. The missions of divers using the Hardsuit™ 2000 are said to be "endless."[1] Oceanworks International makes the Hardsuit™ 2000, and also Hardsuit™ 1000 and 1200, with operational capability to 1,000 and 1,200 feet underwater, respectively. These suits have fans that recirculate oxygen, and a carbon dioxide scrubbing capability. The hulls are aluminum, with digital, voice over communications, vertical and lateral thrusters, pan and tilt video cameras, and color imaging sonar. Oceanworks provides Hardsuits™ for the militaries of the USA, Japan, Turkey, France, Russia and Italy.[2] The suits are also in commercial use in Brazil, Australia, Japan, Canada, the United Kingdom and the Gulf of Mexico. Since 1986, the Hardsuits™ have been used to perform many underwater tasks, including such operations as:

- general intervention
- tunnel and intake inspections
- subsea completions
- construction and repair of structures and control system

---

[1] "Into The Deep," http://www.mediacen.navy.mil/pubs/allhands/jun00/pg14.htm, 2002.

[2] Hardsuit Atmospheric Diving Systems, Oceanworks International, http://www.oceanworks.cc, 2009.

- manifold and tree installation and service [3]

I absolutely am not alleging that either Oceanworks International or their Hardsuit™ 2000, 1200 or 1000 are involved in any way, shape or fashion in any "special access," Top Secret or classified, compartmentalized, secret operation of any sort whatsoever. I simply present the suit and its technological capabilities as an open literature example of the state of the art in modern, deep ocean, marine engineering. In fact the Hardsuit™ is merely the latest iteration of other, similar, deep diving suits that were in use more than 20 years ago, such as the so-called NEWTSUIT that was used in depths up to 750 feet, or the JIM suit, also operable to 2,000 foot depths.[4]

It would be my educated guess, however, that similar technology, likely with heightened capabilities beyond those of the Hardsuit™ atmospheric diving system, is in use at great ocean depths.

From the standpoint of this book's theme, it is important for the reader to understand that from the late-1950s to the mid-1980s one of the practices of the offshore petroleum industry was to put subsea assemblies in dry, one-atmosphere, water tight chambers right on the seafloor. This permitted oil field workers to be sent down to the seafloor in deep water diving bells or small submarines, and to enter the chambers to perform work in a "shirt sleeve" environment down on the seafloor itself.[5] In the mid-1980s, the petroleum industry abandoned this practice, but this sort of capability has obvious implications for construction of non-commercial or clandestine sub-seafloor installations that have openings to the seafloor. Clearly,

---

[3] Ibid.

[4] "Voyagers of the Deep," *Compressed Air* 91, no. 11 (November 1986): 30-39.

[5] Preston Mason, "Evolution of Subsea Well System Technology," *Offshore* 66 issue 7 (July 2006); and M.C. Tate and D.L. Miller, "Design and Application of a Dry Chamber for Subsea Production," in *Underwater Technology: Offshore Petroleum, Proceedings of the International Conference*, Bergen, Norway, 14-16 April 1980, ed. by L. Atteraas, F. Frydenbø, B. Hatlestad, and T. Hopen (New York: Pergamon Press, 1980): 19-32.

chambers of this sort can serve as transfer points for personnel, equipment and supplies to sub-seafloor facilities or bases of the sort discussed later in this book. Interestingly, Lockheed, the well-known aerospace company, developed one of these dry, seafloor, one-atmosphere systems for use in water of 1,200 feet depth, with studies showing that the technology could be extended to depths of 3,000 feet.[6]

Until the development of atmospheric diving suits that could bring divers to depths of 2,000 feet or more within one-atmosphere conditions, divers were actually physically in the water. They therefore lacked protection against the crushing pressure of the deep sea. When working at great depths for prolonged periods, say at 1,000 feet, divers lived in so-called "deck decompression chambers," and traveled to and from work in submersibles or diving bells. It takes several days to decompress from working at a 1,000 feet depth undersea. Clearly, then, working at such depths was highly hazardous.[7]

Liquid Breathing

I raise this point because in the course of my research I have been told by a couple of people that the U.S. Navy actually has a classified deep diving capability that uses a liquid breathing technology, similar to the fluid breathing system used in *The Abyss*, the Hollywood feature film. The plot of that movie revolves around a sunken U.S. Navy submarine, the personnel on an undersea drilling rig, and a

---

[6] Don E. Kash, Irvin L. White, Karl H. Bergey, Michael A. Chartock, Michael D. Devine, R. Leon Leonard, Stepehen N. Salomon, and Harold W. Young, Foreword by Joseph Coates, *Energy Under the Oceans: A Technology Assessment of Outer Continental Shelf Oil and Gas Operations* (Norman, Oklahoma: University of Oklahoma Press, 1973).

[7] "Voyagers of the Deep" op. cit.

team of U.S. Navy divers sent to rescue the sunken submarine.[8]

As a result, I went looking to see if there was any literature on fluid breathing. Intuitively, fluid breathing seems impossible for the mature human organism. Nevertheless, I am aware that our lungs are full of liquid when we are in the fetal stage of our life cycle, inside the womb. So, I searched the Internet and the open scientific literature to see what I could find.

I was surprised to learn that there was some very serious research on fluid breathing done for the U.S. Navy. In the 1970s, Dr. Johannes A. Kylstra conducted a thorough, rigorous series of experiments at Duke University Medical Center for the Office of Naval Research. He did extensive research with lab animals, testing a variety of saline and fluorocarbon compounds, and determined that, with the right gas mixtures and fluid, it is feasible for mammals to breathe fluid and survive. The key factors are the ability of the fluid to move oxygen into the lungs, and to remove carbon dioxide. Dr. Kylstra drew some highly interesting conclusions:

> It should be possible for a healthy man breathing oxygenated FC-80 fluorocarbon to maintain a normal PaCO2 while at rest. This would make possible the rapid escape from disabled submarines at great depth. The use of an emulsion of 1% (by volume) of 2 M NaOH in FC-80 fluorocarbon liquid should permit a liquid breathing diver to perform work requiring a V(O2) of approximately 1 1 STPD/min while maintaining a normal PaCO2.[9]

This is very interesting open literature information. In plain English, Dr. Kylstra is saying that a liquid-breathing diver can breathe a

[8] *The Abyss*, written and directed by James Cameron, 1989. I highly recommend this movie, as it has several themes that relate to some of the content and themes of this book. Having seen the movie, I suspect that James Cameron may be privy to information that the average person does not know. The director's cut is the best version to see, if you can find it.

[9] Johannes A. Kylstra, M.D., *The Feasibility of Liquid Breathing in Man*, U.S. Office of Naval Research contract no. N00014-67-A-0251-0007, Final Technical Report 1 May 1969 – 31 October 1975 (Durham, North Carolina: Duke University Medical Center, 1977).

fluorocarbon and saline solution while performing work underwater at a physiological rate requiring respiration of one liter per minute of oxygen at standard temperature and dry gas pressure, and simultaneously maintaining normal carbon dioxide blood levels. I would be surprised if further classified research was not carried out by the U.S. Navy that pushed the edge of the envelope beyond the parameters that Dr. Kylstra specifies in the abstract of this 1977 report.

The open medical literature also provides other evidence that goes to the feasibility of liquid breathing by humans. A 1996 study published in *The New England Journal of Medicine* detailed research that established that perfluorocarbon liquid was helpful in relieving respiratory distress in premature infants. The abstract of the article states:

> Ten infants received partial liquid ventilation for 24 to 76 hours.... There were no adverse events clearly attributable to partial liquid ventilation. Infants were weaned from partial liquid to gas ventilation without complications. Eight infants survived to 36 weeks corrected gestational age. Partial liquid ventilation leads to clinical improvement and survival in some infants with severe respiratory distress syndrome who are not predicted to survive.[10]

Judging from the data in the open military and medical literature that I have cited here, I believe it is possible that the U.S. Navy has in fact developed and deployed a fluid-breathing diving capability. Technical progress in the world of "special access" programs often greatly exceeds the capabilities of technology in the public realm. I therefore am willing to attach some credence to the stories that have been passed to me.

---

[10] Corinne Lowe Leach, Jay S. Greenspan, S. David Rubenstein, Thomas H. Shaffer, Marla R. Wolfson, J. Craig Jackson, Robert DeLemos and Bradley P. Fuhrman, "Partial Liquid Ventilation With Perflubron In Premature Infants With Severe Respiratory Distress Syndrome," *The New England Journal of Medicine* 335 no. 11 (12 September 1996): 761-767.

U.S. Navy Underwater Construction Teams

Keeping all of the foregoing in mind, it should be remembered that the U.S. Navy has highly trained divers who carry out a wide variety of underwater construction activities. A U.S. Navy website advises that:

> The UCT, comprised of Seabees with specialized dive training, possess underwater repair and construction expertise and are amphibious in nature. They are capable of constructing shallow- and deep-water structures, mooring systems, and underwater instrumentation and also perform light salvage and precision blasting. There are two teams:
>
> UCT-1, based in Little Creek, VA, performs construction and demolition in a combat environment and construction, repair and maintenance of harbor installations, such as piling repair and grouting.
>
> UCT-2, based in Port Hueneme, CA, is responsible for all ocean and waterfront facilities overseen by the Commander of the Pacific Command.[11]

U.S. Navy Mobile Diving and Salvage Units based in Hawaii and Virginia also perform underwater construction missions.[12]

Given everything that I have learned of the U.S. Navy's prospective, manned, undersea bases, the myriad pages of military and corporate documentation that I have examined, what I have been told by a variety of sources over the years, and the revealing illustrations of Walter Koerschner (shown in the next chapter), I conclude that U.S. Navy Underwater Construction Teams have built manned, undersea bases. These are most likely along the Atlantic, Pacific, and Gulf of Mexico coasts of the USA, as well as in other regions such as the Caribbean, Mediterranean, the Indian Ocean, and elsewhere.

Submarines, ROVs and DSRVs

In that vein, the important thing to understand is that modern technology permits ocean engineering projects to be carried out at

---

[11] "Navy Diver: Dive Units," http://www.navy.com/about/navylife/onduty/navydiver/navydiveunits, 2009.

[12] Ibid.

extremely great depths. In recent decades, the development of a variety of very deep diving Remotely Operated Vehicles, submersibles, and submarines has made it feasible to construct underwater facilities and maintain equipment in a deep sea environment.

My first introduction to this Neptunian realm was during the holidays in the last week of 1975. I was hitchhiking on a cold, snowy night from the Scottish Highlands back to London, and happened to thumb a ride with an ex-Royal Navy diver who was then working in the North Sea. He regaled me with tales of his deep sea exploits in the offshore oil fields as we blasted southward down the motorway at speeds upwards of 110 mph. He was driving a really fast little sports car with incredible acceleration. When I glanced at the dashboard I couldn't believe the speeds registering on the speedometer. I was in white knuckle mode the whole time, but he was absolutely unfazed by the snow and ice on the road. It occurred to me that his apparent lack of fear might have been due to ritual libations in his local pub earlier in the evening. I thought of asking him to stop the car and let me out, but at the high speeds he was driving it took me a few miles to formulate the question diplomatically, by which point I had observed that considering everything, including highly probable earlier liquid refreshment and the alarmingly snowy and icy motorway, he was miraculously keeping the car on the road and passing other cars like they were standing still. So I said nothing and tried to hold back my rising sense of panic.

Besides his considerable driving skills, I was impressed by the fact that he went down to the bottom of the North Sea and worked on pipelines, well heads and other subsea, oil field infrastructure, and actually performed heavy construction work deep underwater. The fact that he routinely welded undersea was an eye-opener for me. Prior to that late-night, race car dash through the snow from

Scotland to London, I had had no idea that it was possible to do heavy, industrial construction down on the seabed itself.

I also saw first-hand the type of person who goes down hundreds, or even thousands, of feet underwater, and does heavy construction work in a deep sea environment. They're not wired for fear like most people! It takes a certain fearless, daredevil attitude even to entertain the idea of diving that deep, coupled with a rare set of physical skills – strength, hand to eye coordination, agility, and catlike reflexes. The margin for error is very slim when you are hundreds or thousands of feet underwater. The deep sea is very unforgiving of mistakes.

ROVs

Because of the perils of putting men directly out in the deep sea and exposing them to the dangers of the crushing depths and freezing temperatures of the seafloor, a new generation of machine has been developed with the capability to dive miles below the surface and carry out heavy work underwater. Remotely Operated Vehicles (ROVs) are in wide use by marine industry and naval military agencies around the world.

The data in the open literature show that ROVs can operate at depths in the neighborhood of four miles and carry out a wide range of scientific, industrial and military activities. The U.S. Navy's CURV III system is representative of the state-of-the-art:

> The CURV III is a 12,600-pound Remotely Operated Vehicle (ROV) that is designed to meet the Navy's deep-water salvage requirements down to a maximum depth of 20,000 feet of seawater.... For special operations, the ROV can accommodate customized, skid-mounted tool packages. These packages can include, but are not limited to, trenchers, specialized salvage tools, instrument packages or other mission-oriented equipment.[13]

The whole system comes with sophisticated fiber optics, sonar, black

[13] "CURV III," http://www.supsalv.org/00c2_curvIIIRov.asp, 2009.

Illustration 5-1: CURV III, U.S. Navy ROV for deep sea operations. *Credit:* U.S. Navy, 2009.

and white and color television, still camera, and digital data and communications network.

Private industry also operates hundreds of ROVs all over the world. These also have the ability to work at great depth in the offshore oil industry, in support of scientific research, in salvage operations, and other applications.[14]

Mystic and Avalon

Which brings me to the curious matter of the U.S. Navy's so-called Deep Submergence Rescue Vehicles (DSRV), of which only

---

[14] "Voyagers of the Deep," *Compressed Air* 91 no. 11 (November 1996): 30-39; also see the Remore 6000 ROV from Phoenix International, http://phnx-international.com/Remora6000 ROV.htm, 2009.

two were reportedly manufactured and deployed, both of which are now supposedly retired from service. The DSRVs were named *Mystic* and *Avalon*, and were ostensibly developed and deployed as a system to rescue the crews of sunken submarines, in a deep sea environment.

The *Mystic* and *Avalon* were built by the Lockheed Missiles and Space Company in Sunnyvale, California. It is interesting that the major shipyards that normally build the U.S. Navy's submarines played no role in constructing the DSRVs. Rather, a leading American aerospace company carried out the R&D and the construction. Notice that Lockheed was also mentioned above as a manufacturer of manned, one-atmosphere, "dry chambers" used to put men on the seafloor in a shirtsleeve environment to work on undersea oilfield equipment. The question naturally arises as to what Lockheed may be up to in the deep sea environment. According to the Navy, the DSRVs displaced 38 tons, were 49 feet long, 8 feet in diameter, moved at 4 knots, had a crew of two along with two rescue personnel, could carry twenty-four passengers and had a rated depth of 5,000 feet.[15] First launched in 1970, the U.S. Navy says that:

> Deep Submergence Rescue Vehicles perform rescue operations on submerged, disabled submarines of the U.S. Navy or foreign navies.... At the accident site, the DSRV works with either a "mother" ship or "mother" submarine. The DSRV dives... and attaches to the disabled submarine's hatch. DSRVs can embark up to 24 personnel for transfer to the "mother" vessel.[16]

Over the years, the U.S. Navy has had joint exercises with other navies to practice submarine rescue operations. DSRVs can be transported by surface ship or transport aircraft. Moreover, as of 2002, eight U.S. Navy submarines were configured to provide

[15] "Deep Submergence Rescue Vehicle – DSRV," http://www.chinfo.navy.mil/navypalib/factfile/ships/ship-dsrv.html, 2002.

[16] Ibid.

piggyback-style transport of DSRVs. So you see, the capability existed to transfer dozens of crew members from a stricken submarine and take them on board another submarine – while still at sea. This procedure was explained by the *Mystic's* Lead Chief Petty Officer, Todd Litke:

> To man our DSRV, break away from the mother submarine, locate and mate to a disabled submarine, transfer all personnel, and come back should take only four to five hours.[17]

Though Mr. Litke does not say so, maybe because he does not even know so, the same procedure can also be carried out in reverse, to transfer personnel to a submerged submarine, or to mate with an airlock on the entrance to an undersea base, for instance. I suspect that secret transfer of personnel to clandestine, undersea bases may have been one of the operations carried out by the DSRVs.

Illustration 5-2: The DSRV *Mystic* being mated with a mother submarine while in port. *Source:* U.S. Navy.

[17] "Submarine Rescue," http://www.navy.mil/navydata/cno/n87/usw/issue_15/submarine_rescue.html, 2009.

The NR-1

In the 1960s, the U.S. Navy also developed a very unique, nuclear powered submarine called the NR-1, that has been widely used for military intelligence, marine research and even undersea archeology. The NR-1 was loaded with scientific and engineering instruments and equipment. According to the Navy:

> The NR 1 performs underwater search and recovery, oceanographic research missions and installation and maintenance of underwater equipment, to a depth of over half a mile. Its features include extendable bottoming wheels, three viewing ports, exterior lighting and television and still cameras for color photographic studies, an object recovery claw, a manipulator that can be fitted with various gripping and cutting tools and a work basket that can be used in conjunction with the manipulator to deposit or recover items in the sea.[18]

The NR-1 was especially valuable to the Navy because its nuclear power plant permitted it to linger for long periods of time on the seabed without needing to surface. The NR-1 was retired in late 2008.

Curiously, the NR-1 was never commissioned by the U.S. Navy. Nor did it have a name like other submarines, say, the USS Something-or-Other. It was known by its serial number. It could dive to 3,000 feet with a crew of eleven. Other unusual features were alcohol-filled wheels that permitted it to actually roll along the sea floor, like a motor vehicle. The NR-1 was not a fast submarine, moving at about 3 knots, but it could linger indefinitely at great depth. Other unique features included three viewing portholes, advanced sonar systems, and multiple cameras that it could use to image the ocean floor. It was used to search for "wrecked and sunken naval aircraft," and right up to the end of its career was being used

[18] "Deep Submergence Craft – NR 1," United States Navy Fact File, http://usmilitary.about.com/library/milinfol/navyfacts/bldeepsubcraft.htm.

Illustration 5-3: The Deep Submergence Vessel NR-1 on the surface in Port Canaveral. *Source:* U.S. Navy.

for "highly classified military missions."[19] Many of the NR-1's missions remain classified to this day.[20]

The Sea Cliff, Trieste-II, USS Dolphin, Seawolf, Halibut, Parche and Jimmy Carter

In addition to the NR-1, many other submarines and submersibles were developed with a deep diving, special operations capability. Some of them are publicly known. Presumably others are not. I will briefly mention just a few from the last 40 years that are known. The list is representative, not exhaustive

The diesel-electric USS *Dolphin* was deployed in 1968 and deactivated in 2006. It carried a crew of 46 enlisted men, 5 officers and up to 5 scientists. The *Dolphin* carried out many scientific and

[19] Andrew Scutro, "Deep-diving NR-1 wraps up its 40-year career," http://www.navytimes.com/news/2008/11/navy_nr1retires_113008w/, 2009.

[20] Lee Vyborny and Don Davis, *Dark Waters: An Insider's Account of the NR-1, The Cold War's Undercover Nuclear Sub* (New York: New American Library, 2003).

military missions, including many that were secret. It was one of the most unique submarines in the world, with the capability of diving below 3,000 feet.[21] Indeed, one authority cited a depth of 4,000 feet.[22]

The USS *Parche*, now retired, had the capability to insert divers directly into the water while submerged.[23] The USS *Jimmy Carter*, which replaced the *Parche*, also has this capability.[24] The *Jimmy Carter* has extraordinary surveillance capabilities. While my research suggests that all submarines in the U.S. Navy fleet can be modified to deploy divers while submerged, the *Jimmy Carter* is exceptionally advanced in this regard, carrying a large complement of special forces divers that it can deploy directly to the undersea environment at great depth. It is also rumored to have an extremely deep, diving capability.[25] The USS *Seawolf* and USS *Halibut* were also used in past decades for a variety of sensitive missions that included tapping undersea cables and putting divers out undersea at great depth.[26]

The *Trieste II* was an extremely deep diving submersible that was an extensively modified and improved version of the original *Trieste*

---

[21] "Research Submarine – USS Dolphin (AGSS 555)," http://usmilitary.about.com/library/milinfo/navyfacts/blresearchsub.htm, 2002; and Steve Liewer, "Tight budget forces Navy to put squeeze on sub," http://www.signonsandiego.com/news/military/20060828-9999-1n28dolphin.html, 28 August 2006.

[22] Will Forman, *The History of American Deep Submersible Operations 1775-1995* (Flagstaff, Arizona: Best Publishing Company, 1999).

[23] Robert A. Hamilton, "Super-Secret Sub Goes Out Of Service," New London Day, http://www.chinfo.navy.mil/navpalib/.www/rhumblines/rhumblines565.doc, 20 October 2004.

[24] "USS Jimmy Carter SSN-23," http://www.submarinehistory.com/JimmyCarter.html, 2009. Also see "Jimmy Carter: Super Spy?," http://www.defensetech.org/archives/001397.html, 2009.

[25] Joe Buff, "USS Jimmy Carter: SSN-23," http://www.military.com/NewContent/0,13190,Buff_060704,00.html, 2004.

[26] Sherry Sontag and Christopher Drew, with Annette Lawrence Drew, *Blind Man's Bluff: The Untold Story of American Submarine Espionage* (New York: Public Affairs, 1998). And John Piña Craven, *The Silent War: The Cold War Battle Beneath The Sea* (New York: Touchstone, Simon & Schuster, 2001).

bathyscaph. It was the original *Trieste* that in 1960 took a two-man crew to the deepest manned dive ever attempted – the depths of the Mariana Trench, seven miles below the ocean's surface. In the late 1950s the U.S. Navy purchased the *Trieste* from its private owners and then extensively modified it over the years, making many classified dives to very great depths in oceans and seas all over the world for purposes of salvage, research, reconnaissance, search and other missions. Most of the missions cited in the open literature ranged in depth between 4,000 and 20,000 feet. The last incarnation of the *Trieste* was retired in 1984 after a very active service life.[27]

I should also briefly mention the *Sea Cliff*, which was the deep diving submersible that replaced the *Trieste* when the latter was retired. Carrying a crew of 3, the *Sea Cliff* was able to operate at depths of 20,000 feet while being more maneuverable than the *Trieste*.[28]

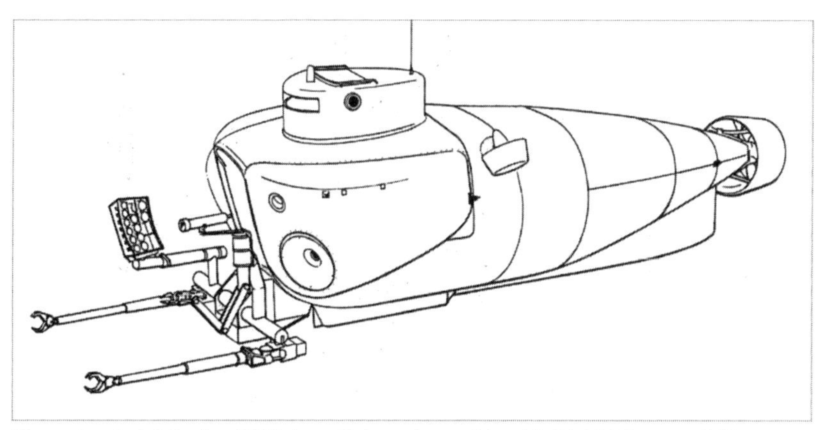

Illustration 5-4: The Sea Cliff, deep diving submersible with 20,000 foot capability. *Source:* NOAA, U.S. Commerce Department, 1983.

[27] Craven, 2001; Forman, 1999; and "History of the Bathyscaph Trieste," http://www.bathyscaphtrieste.com, 2009.

[28] John Freund, "Sea Cliff (20,000 Foot Modification Project," *Eleventh Meeting of the United States-Japan Cooperative Program in Natural Resources (UJNR)*, Panel on Marine Facilities, National Oceanic and Atmospheric Administration, U.S. Department of Commerce, March 1983.

Of course, there is a great deal more to be said about all of these submersibles and submarines, but that is not my purpose here. Those who are interested in greater detail are advised to consult the source material in the footnotes in this chapter. The topic is a fascinating one and the more you read and research, the more connections you make, and the greater the appreciation you develop for the technical capabilities of the Navy and industry for operating in the deep sea environment.

More To The Story

There is likely to be a great deal more to this story of deep sea operations than the U.S. Navy has presented to the public. As I have documented, the U.S. Navy and industry have had a wide variety of submersibles and submarines able to work undersea at very great depths. They can even deliver manned work parties to the seafloor hundreds and even thousands of feet underwater. This capability has existed for about 40 to 50 years. What is much less well known is that since the 1960s, the U.S. Navy has had a very advanced capability to work miles below the surface of the sea, on and even below the seafloor. My research indicates a strong likelihood that this capability includes putting manned vessels and work parties on the seafloor and far below – way down inside the underlying bedrock in manned, undersea bases in mid-ocean, or off the coasts of continents and islands anywhere in the world. Certainly this includes the coasts of North America, Europe, and the nearby islands. Specifically, the Pacific, Atlantic, and Gulf of Mexico coasts of the United States, the Caribbean and Aleutian islands, the Atlantic and Mediterranean coasts of Europe (including the Baltic and North Seas, the Bay of Biscay, and English Channel), and many mid-sea and mid-ocean archipelagos such as Bermuda, Hawaii, Guam, Samoa, and the

Canary, Azorean, Falkland, Faro, Shetland and Balearic Islands, the Greater and Lesser Antilles, and more.

Under the direction of John Piña Craven, the Chief Scientist of the U.S. Navy's Special Projects Office from 1958 to 1970, the Navy dramatically extended its capability to operate at extreme depth in a deep sea environment. After the sinking of the nuclear submarine, USS *Thresher*, in 8,000 feet of water in 1963, the U.S. Navy wanted a feasible means of rescuing sunken submariners from great depth. John Piña Craven was accordingly tasked by the Navy with developing the technology to do just that.

He then led a crash project to develop a capability to operate at depths of as much as 20,000 feet underwater, which included development and deployment of the DSRVs, *Mystic* and *Avalon*, and the nuclear NR-1 submarine, as well as the *Trieste* and *Sea Cliff*, all of which I have mentioned above. However, the fact that the *Thresher's* hull imploded catastrophically at a depth of about 2,000 feet meant that survival of the crew at greater depths was virtually impossible. So why the drive to go 20,000 feet deep, if no one could feasibly survive such a crushing depth and await rescue?

I believe there were two reasons, both of them hinted at by Craven himself, in his public work and also, less directly, by the work of others. Both reasons are connected to some of the most deeply classified missions the U.S. Navy has ever carried out. To wit, I believe that Craven quite possibly: 1) headed up search and retrieval of crashed, sunken UFOs from the deep sea, and 2) set in motion construction of a super-secure system of very deep, manned installations located down in the seabed itself, far below the seafloor, well out to sea.

Permit me to explain. In 1964, John Piña Craven was put in charge of a special, independent program within the U.S. Navy called the Deep Submergence Systems Project (DSSP). This was only the second instance of such an independent program being developed within the United States Navy's organizational structure. Shortly thereafter, in 1965 Craven was given a high-level intelligence briefing by a Naval Intelligence officer and initiated into the hermetic world of highly classified, compartmentalized programs. Briefly, he was informed of the existence of an extremely sensitive program, code-named Sand Dollar. The program's objective was to retrieve militarily sensitive hardware and other items with national security importance from the seafloor of the continental shelf and the deep marine environment. Judging from Craven's narrative in his memoirs, Sand Dollar was buried within the structure of yet another secret program, itself hidden within the Polaris submarine program. He was further informed that, in the Top Secret realm of compartmentalized, special access programs, a "hierarchy of projects" existed. First, a publicly known project or program would provide an umbrella, under which secret projects could be carried out. The "host" project would thus provide concealment, serving to deflect detection or scrutiny from more sensitive activities carried out under its aegis. The DSSP, Craven was told, could serve in such a capacity for other projects. A further requirement of involvement in the world of nested, tightly compartmentalized secret operations was that the individuals involved could never disclose their participation, or the nature of their work to anyone without a need to know. This included their family members, friends, associates or co-workers.[29]

Secret, Undersea Bases and Retrieval of Crashed UFOs

O.K., O.K., I can hear you thinking it now. You were with me so far, but now that I've broached the issue of crashed UFOs, you're starting to wonder where I'm going with the discussion. The short

[29] John Piña Craven, *The Silent War: The Cold War Battle Beneath The Sea* (New York: First Touchstone Edition, Simon & Schuster, 2002).

answer is that I am headed directly into the question of one of the highest security policy areas on this planet: the question of so-called UFOs and extraterrestrial and/or ultradimensional beings, civilizations and their technology, as surveilled, interrogated, recovered and sequestered in great secrecy by agencies and projects with classified names.

Perhaps even by a project named Sand Dollar. To start with, Craven begins by saying that the name of the Sand Dollar program is classified "to this day" (writing in 2002), so that he is bound by his security oath not to reveal its name – but he then cavalierly reveals the name anyway, proffering the explanation that it doesn't matter if he does so, because the name is already in the public domain. This "explanation" of his raises obvious questions which he never resolves. A few pages later he explains that Sand Dollar attached the "highest priority" to the recovery of sunken nuclear weapons from the world's oceans. Interestingly, he does not say that the recovery of sunken nuclear weapons was the only priority. He then states:

> I was shown an inventory itemizing the items known to be on the seabed and a map of their distribution throughout the world.[30]

By implication, the obvious inference would be that he is principally referring to nuclear weapons. However, the word, "items," is not modified, and admits of the possibility that other "items" of interest were also inventoried for retrieval. On the next page he discusses the great value of the DSSP as an umbrella bureaucratic structure for a crucial, undersea national security program, and his discovery that "items" of interest were scattered all over the seafloor worldwide, not just in the infamous Bermuda Triangle. In his words:

> The salvageable items of interest were spread around the globe.... There were plenty of interesting objects on ocean floors worldwide, not just in the

[30] Ibid.

> Bermuda Triangle, and there was an equal distribution of these objects between those that were lying on the continental shelf and those in the deep ocean. There was a skew in favor of the continental shelf because of shallow water, rocks and shoals, but much of that hardware had either been salvaged or scattered by storms.[31]

So, as of the mid-1960s there were "plenty of interesting objects" strewn across the seafloor all over the world. And much of it on the continental shelves had been already been "salvaged or scattered by storms." So what are we talking about here? Nuclear weapons for one, presumably mostly Russian and American, but he is clearly talking about more. Maybe Spanish galleons? Sunken Roman galleys? German U-boats? Something else, maybe even more exotic? And who, or what, had already salvaged much of that unspecified "hardware"? And if it had already been "salvaged" why would it still be on a super secret U.S. Navy list of items of salvageable interest? Craven's choice of language is both revealing and ambiguous at the same time, revealing that the Navy had a global inventory of objects of high interest strewn across the bottom of the planet's oceans, but ambiguous enough to conceal the exact nature of what these objects were. Elsewhere in his book, he discloses some of the efforts to find and retrieve both American and Russian hardware, submarines and weaponry, from the ocean floor. All of the cases he mentions have been discussed elsewhere in other organs and forums by the main-stream news media in previous decades. Clearly there were other, very sensitive, undersea missions that Craven does not directly discuss. But he does drop some hints.

Beside revealing early on that the Sand Dollar and DSSP programs could serve as "host" programs for other secret projects hidden under their cover, he makes other revealing comments. In discussing top secret missions carried out by the submarines, *Seawolf*

---

[31] Ibid.

and *Halibut*, he writes that, paradoxically, there was no need for "maximum security" for the submarines themselves because:

> ...(T)hey only transported top secret equipment and teams of industry specialists who would carry out unknown missions in unknown parts of the seas. The performers and the performances were invisible.[32]

And he adds that *Halibut* had a DSRV "mock-up" mounted on its deck which the crew was evidently led to believe was a "temporary expedient," while *Seawolf* had been cut in half, in part for refueling her reactor, but also to add other unspecified equipment and men. All to carry out "unknown missions" in "unknown parts of the seas" using "top secret equipment" and teams of "industry specialists," all while the missions and "performers" remained "invisible."

There is no doubt that many of these mysterious missions involved undersea communications cable surveillance missions, and probably the construction of undersea bases, which is extensively discussed in this book, but I also believe that another part of the answer to this mysterious matter has been provided by the ground-breaking work of the UFO researcher, Ryan S. Wood. In his excellent book, *Majic Eyes Only*,[33] Wood rigorously documents that military agencies all over the world have undertaken scores of retrieval operations of downed or crashed UFOs, as well as retrieval of materials from damaged UFOs. Interestingly, some of these operations have included American military agencies and personnel, including the U.S. Navy. Of special note for this discussion is the fact that a number of these operations took place in coastal areas or actually out in the open sea. Wood mentions two that stand out for me: 1) Shortly after the so-called Los Angeles Air Raid on 25 February 1942, at the height of WW-II, when coastal antiaircraft

---

[32] Ibid.

[33] Ryan S. Wood, *Majic Eyes Only* (Colorado, USA: Wood Enterprises, 2005).

batteries opened up a furious, overnight barrage against a mystery aerial object hovering over the Los Angeles metropolitan area, Naval Intelligence reported the off-the-coast recovery by naval salvage of an airplane that was "not earthly" and a subsequent report of the incident to President Roosevelt; and 2) in June 1973, the U.S. Navy reportedly picked up on radar, shot down and retrieved a UFO from the Pacific Ocean, with the aid of the Glomar Explorer vessel.[34]

In my previous book, *Underwater and Underground Bases*,[35] I devote several pages to the mysterious, secretive activities of the late Howard Hughes' Global Marine Company which launched the so-called Glomar vessels to carry out deep sea salvage and drilling missions for the CIA, and ultimately other agencies. John Piña Craven mentions the Glomar Challenger in his book, but gives the reader the impression that he was out of the loop with regard to its mission and activities.

I am not a bona fide statistics expert like John Piña Craven, but I would give you better than even odds that the DSSP and Sand Dollar programs that Craven worked on carried out at least a few UFO retrievals of precisely the sort that Ryan Wood has documented. Moreover, that a fair number of the items of interest that U.S. Naval Intelligence wanted him to retrieve from the seafloor worldwide were of a highly exotic, possibly even extraterrestrial nature. It is my hope that anyone reading this book with knowledge of ocean retrievals by the U.S. Navy or CIA of crashed or downed UFOs will contact Ryan Wood.[36] For years, ever since I began my research in the early 1990s, people have been telling me that the U.S. Navy is a lead agency in the highly secretive realm of extraterrestrial

---

[34] Ibid.

[35] Richard Sauder, *Underwater and Underground Bases* (Kempton, Illinois: Adventures Unlimited Press, 2001).

[36] Wood's website address is: http://www.majesticdocuments.com.

and UFO-related technology, research and intelligence. It would not surprise me at all if some of these activities were to have been secreted away in such a tightly compartmentalized project as the DSSP program.

Undersea Submarine Tunnels

Since the very beginning of my research, people have also been telling me of massive tunnels off both coasts of the continental USA. Ocean-going submarines were said to enter these, well out at sea, then travel to inland, underground submarine bases, where they would dock in great secrecy.

At first I was skeptical of the stories, but I give them more credence now. I was told these tunnels exist in the Long Beach, California area. Other candidate sites are the Puget Sound region of the Pacific Northwest, the San Diego and Channel Islands region of southern California, the Monterey Bay, California area, the Hawaiian archipelago, and off the East Coast of North America, at places like Kings Bay, Georgia, Mayport, Florida, Norfolk, Virginia, and offshore from Massachusetts. My research also points to candidate sites for similar facilities in Great Britain, on the coast of northwest Scotland, and the North Sea coast.

Both the Chinese and Russians have made such underwater tunnels and bases for their military submarines. In *Blind Man's Bluff*, Sherry Sontag and Christopher Drew reveal that already in the late Soviet period, the Russians were making underwater tunnels for their enormous Typhoon class, nuclear-missile-firing submarines on the Kola Peninsula, about 150 miles from Murmansk.[37] And in 2008, author Gordon Thomas revealed that the Chinese had built a massive, underground naval base that can accommodate up to 20 military submarines inside Hainan Island. It includes as many as 11

[37] Sherry Sontag and Christopher Drew, *Blind Man's Bluff.*

separate underwater tunnels through which Chinese nuclear-missile-firing submarines can travel, on their way to and from the base, submerged the entire time, coming and going from the open sea.[38] If the Russians and Chinese can make underwater tunnels for their submarines, then the U.S. Navy can, too. No doubt, the stories I have heard are substantially true. Elsewhere in this book I show the illustrations of Walter Koerschner, tasked by the U.S. Navy circa 1968 with drawing up a series of colorful illustrations for the Navy's undersea base R&D effort. In those illustrations, you can see quite clearly that more than 40 years ago, the U.S. Navy was planning to construct underwater tunnels through which submarines could travel. Thus, the circumstantial evidence points towards the U.S. Navy having at least the capabilities of the Russians and Chinese.

Interestingly, one expert source with whom I spoke casually mentioned, in an off-hand way, that one innovative way to potentially make such submarine tunnel systems and bases would be to make use of large, empty, already-existing lava tubes from extinct volcanoes of the sort found in volcanic islands. Using large, high-powered reams pulled through the lava tubes, the tubes could be progressively enlarged until they were suitable for the passage of deep sea submarines that could enter them at sea, below the water line, then travel through the mechanically enlarged lava tube to an onshore, underground submarine base, perhaps in a reconfigured, empty magma chamber located on the interior of an extinct volcano. There are many ancient volcanic islands in the world's oceans and seas that could potentially be engineered in this manner for clandestine submarine bases. This could also include submerged, extinct volcanic sea mounts with empty lava tubes and empty magma

---

[38] Gordon Thomas, *Secret Wars: One Hundred years of British Intelligence Inside MI5 and MI6* (New York: Thomas Dunne Books, St. Martins Press, 2009); and Gordon Thomas, "Is Chinese Regime Preparing for Nuclear War?," http://en.epochtimes.com/news/8-5-20/70796.html, 20 May 2008.

chambers that could be converted into clandestine undersea bases. My source did not tell me that this has been done, only that it hypothetically could be done. However, I can tell you that I have read through many thousands of pages of military, marine, geologic and marine engineering literature, and from the technical and engineering capabilities that I have seen, the U.S. military and private industry certainly have the money, technology, machinery, technical and engineering expertise, and trained personnel to carry out construction operations of this sort. They can do this hundreds or thousands of feet underwater – in volcanic sea mounts, as well as under many other undersea, geological conditions. It is likely that the military agencies and large engineering corporations of other major nations have similar engineering prowess. The Chinese example cited by Gordon Thomas is illustrative of the state of art in modern marine military engineering. We live in a science fiction world. It is time that we adjust our thinking, our perceptions, and our world view accordingly.

Of course, the inevitable question arises as to how undersea bases can be supplied with oxygen for the crew to breathe. The counterintuitive answer for air breathing humans is that the sea is full of water! This matters because the chemical composition of water molecules consists of a 2:1 ratio of hydrogen to oxygen. By means of electrolysis, the oxygen and hydrogen atoms that comprise the water molecule are separated, yielding pure oxygen. The oxygen is retained for breathing. The hydrogen can be used as a clean burning fuel or discarded. This technology was perfected by private industry a half century ago and is in common use today. Since 1959 the Treadwell Corporation has provided the U.S. Navy with electrolytic oxygen generation systems to provide breathing air for the crews of its nuclear submarines.[39] The same technology can just as easily be used to provide oxygen in enclosed, manned undersea bases. The U.S. Navy also uses air scrubbers that remove carbon dioxide and other contaminants from the air on submerged submarines.[40] Clearly, the same technology can be employed in undersea bases to purify the air and render it breathable.

I now accept that construction of manned, clandestine undersea bases has taken place, far down in the bedrock beneath the seafloor. The U.S. Navy, private industry, and other organizations and agencies have built an unknown number of these facilities around the world. John Piña Craven himself left the door open for this type of speculation, and more, when he said in 2002, with respect to the DSRVs *Mystic* and *Avalon*:

> The DSRV's technical sophistication and complexity surpassed the space vehicles of that day and perhaps of today. This was a vehicle ahead of its time, with versatilities still unpublicized, and the potential to conduct missions not yet publicly articulated.[41]

The DSRVs' submarine rescue capability is, of course, well known and has been publicly reported and discussed. But here Craven is saying that the DSRVs have other, "unpublicized versatilities" and the ability to carry out missions that haven't yet been "publicly articulated." Can he make it any clearer? He is saying plainly that the DSRVs do something else, in addition to rescuing sunken submariners; that there is more to what the DSRVs do and the manner in which they do it, than has been publicly revealed. I have made a case, based on a variety of circumstantial evidence and information leaked to me by a wide array of individuals, that those

---

[39] "OGP (Oxygen Generation Plant) Electrolysis Module," http://www.treadwellcorp.com/ogpem.htm, 2009. Also see http://www.treadwellcorp.com/history.htm, 2009.

[40] "EADS Space Transportation - Air Contamination Control Devices," http://www.naval-technology.com/contractors/hvac/dornier, 2009.

[41] John Piña Craven, *The Silent War*.

other missions plausibly include traveling to and from clandestine manned bases, buried in the bedrock, beneath the ocean floor, out on the continental shelves, as well as in mid-ocean.

In other words, I postulate that the DSRVs have a submarine shuttle function that includes clandestine crew transfers to and from secret undersea bases. I believe that this is probably what Dr. Craven is broadly hinting at. But why then does he compare the DSRVs to space vehicles? They were manufactured by Lockheed, a major aerospace company, that is true enough, so what "missions not yet publicly articulated" did the DSRVs actually carry out? What is John Craven really talking about? We don't know for sure and he isn't saying, so we are left guessing. It is clear, however, that these submersibles had other missions that have never been publicly revealed.

Chapter 6:

Way Down Beneath the Moonless Mountains

In "Letters From the Underground: An Alien Carousel," an article that initially appeared in the November-December 2003 issue of the former *UFO Magazine* published out of Leeds, U.K., I made a case for the existence of clandestine, sub-bottom, manned, undersea bases. I continue to actively research this possibility and amazingly, new information continues to surface, bit by bit.

On more than one occasion, I spoke about my interest in this topic with the late editor of that magazine, Graham Birdsall. I clearly remember him telling me of stories that had been related to him of unknown aerial objects and craft seen entering and leaving the sea. In addition, the writings of several other authors such as Linda Moulton Howe, Timothy Good and Jorge Martín have also raised the issue of possible alien undersea bases in far-flung locales as diverse as the Caribbean Sea and the Pacific Ocean. My own recent work has explored the likelihood of clandestine man-made undersea bases.

The Magnificent Artwork Of Walter Koerschner

Notwithstanding all of that, I was taken completely unawares when, in the autumn of 2003, I was unexpectedly contacted by a gentleman who chanced to hear me interviewed on the popular, late night radio talk show, *Coast to Coast AM*, with host George Noory.

This is the biggest of the nighttime radio talk shows in the United States, and millions of people tune in every night. But to my pleasant surprise, this particular listener had a most unexpected professional background. It did not take me long at all to appreciate the unique talent and expertise of Walter Koerschner. He had been attached as an illustrator to the U.S. Navy's Rock-Site team back in the 1960s, at the Navy's China Lake, California Weapons Center. As those familiar with my recent work will remember, this was the time and place when the United States Navy began to develop detailed plans for siting large, technologically complex, deeply buried manned bases beneath the sea floor, in mid-ocean. I provide complete documentation for this in my 2001 book, *Underwater and Underground Bases*.

So you can imagine my interest when Mr. Koerschner mentioned to me that he still had some of the original illustrations he had done for this project. And when he offered to send them to me for publication I immediately accepted his kind offer. It is directly due to his considerable artistic talent and spirit of generosity that we all are now privy to a birds-eye view into the designs of the Rock-Site planning group circa 1968. What I will do is simply present his wonderful graphic illustrations for you to view, along with some descriptive commentary and analysis. The old saw that "a picture is worth a thousand words" certainly holds true in this case. Many things that are described verbally in the Rock-Site documentation become so much easier to understand when you can see them depicted visually.

The Stanford Research Institute Connection

Before diving into Walter Koerschner's undersea pictures, however, I want to add that at about the same time he contacted me, my ongoing archival and documentary research uncovered a document, also from 1968, that explicitly discussed the construction of dozens of deep undersea bases. Two researchers affiliated with the Stanford Research Institute in Menlo Park, California, published a study entitled, "Feasibility of Manned In-Bottom Bases."[1] I want to reproduce the complete abstract from their study because it is so revealing of what the military-industrial complex was planning at that time, and shows so clearly what was considered within technological reach – as of four decades ago. Keep in mind that in the intervening years technology has only grown more powerful and more sophisticated. Let's look at the abstract:

Abstract

The construction of thirty manned in-bottom bases within the ocean floors is technically and economically feasible. However, it will be necessary to establish some successive types of experimental facilities before a full construction program can be started. This could take 15 years.

The major technology for a land-linked station in-bottom is established now; only adaptations are needed. The remaining experimental phases will require further development of equipment and techniques applicable to remote sea access. There are useful assignments for a succession of three experimental stations other than advancing in-bottom construction techniques. Science and engineering concerned with the oceans and their resources will be furthered and military tests of undersea base functions complementing deeper operations can be accomplished. The costs of the experimental phases, called here a demonstration program, can be surprisingly modest: approximately one half-billion dollars, spent over 15 years.

A distinction between in-bottom and on-bottom facilities is made in the numbers of men enclosed and the depth of water, wherein areas of one atmosphere space can be created in-bottom at such low cost the ingress-egress system can be amortized if the space required is reasonably large. Economics thus can dictate choice between the two types; even so, some on-bottom facilities will be needed to aid the construction of remote in-bottom facilities.

Presently, establishing an in-bottom facility and building upon this will present fewer technical difficulties than do the submersibles which would

[1] T.G. Warfield and L.R. Parkinson (Stanford Research Institute, Menlo Park, California), *Feasibility of Manned In-Bottom Bases*, AIAA Paper No. 68-479, 3rd Marine Systems and ASW Meeting, San Diego, California, 29 April – 1 May 1968.

> support it and use it. Subsequent to the completion of the third phase of a demonstration program, which would be a remote, deep water station, and the evaluation of it, a multiple base program could be implemented. The cost of such a base program would be about $2.7 billion for construction of a number of bases (assumed at 30).[2]

Well, that throws everything into sharper perspective. From the outset, it appears that the plan was to make dozens of undersea bases. The document itself makes clear that money would not be a problem. The figure of $2.7 billion – to build 30 in-bottom, undersea bases – is relative peanuts, even by the black budget standards of the late 1960s. In the world of today, of course, funding ought to be easier still, considering the hundreds of billions of dollars, if not trillions, that have gone conspicuously missing at the Pentagon. Interestingly, this particular plan, which makes clear and unambiguous reference to sharing of information by the Naval Underwater Warfare Center at China Lake, also mentions the desirability of making the planned, in-bottom undersea bases to be "multipurpose." That is, these facilities could serve industrial (hard-rock mining and oil drilling) and scientific (geological, seismic, oceanographic) purposes, in addition to military objectives. This raises the possibility that the military might occupy and use such a base, under cover of a scientific or, even better, an industrial operation, which itself could conceivably be clandestine.

I find further significance in the communication between the team at China Lake and the team at Stanford Research Institute. For many years, SRI has played a key role in the military-industrial-espionage complex. It is a place where cutting-edge military policy can transform into clandestine projects – well out of the public eye. Perhaps the most salient example of this is the now well-known psychic military espionage program that fielded a team of so-called

[2] Ibid.

remote viewers in the last quarter of the 20th Century. The genesis of this program was at SRI back in the 1970s, when the early work and training took place. Later the project moved to Fort Meade, Maryland and much of the training shifted to the Monroe Institute, in Virginia. But the initial impetus of this highly clandestine, psychic espionage program came out of SRI.[3] Has something similar happened in the case of the manned, in-bottom, deep undersea base plans? The question naturally arises, given the SRI affiliation of the authors, and their contact with the undersea bases team at China Lake.

Further documentation, indicating the strength of the U.S. Navy's interest in constructing manned, in-bottom bases deep beneath the sea, also surfaced in 1969 in the form of an interview given to the journal *Astronautics & Aeronautics*, by William B. McLean, the inventor of the Sidewinder air-to-air missile and former Technical Director of the China Lake Naval Ordnance Test Center (NOTS), from 1954 to 1967.[4] In 1967 Dr. McLean left China Lake to become Technical Director of the U.S. Naval Undersea Warfare Center in San Diego, California and held that post until his early retirement in 1974. This is significant information, because in both of these positions Dr. McLean evidently had supervisory authority over the Navy's plans and interest in designing and constructing undersea bases. Moreover, the time frame of the mid-to-late 1960s is precisely the time when these plans and interests were being actively developed. So his remarks to then Editor-in-Chief, John Newbauer, of *Astronautics and Aeronautics* carry particular weight.

---

[3] Jim Schnabel, *Remote Viewers: The Secret History of America's Psychic Spies* (New York: Dell Publishing, 1997).

[4] "A Bedrock View of Ocean Engineering," Interview of William B. McLean by A/A Editor-in-Chief John Newbauer, *Astronautics & Aeronautics* (April 1969): 30-36.

Of greatest relevance for this research is his acknowledgment that making an undersea base down in the rock that underlies the ocean floor:

> ... immediately frees you from limitations of size and weight, once you get established in the undersea environment – below the water and the rock. Existing mining techniques would allow you to extend your environment without limits. These techniques have already produced thousands of miles of undersea mining corridors, extending out from a coastal entrance, and in one instance I know of, from the ocean bed.[5]

Dr. McLean goes on to say that where undersea bases are concerned, "Mining techniques can drill as big a cavity as necessary," and that he sees "...no real size limit on the cavity you can build- at costs not greatly exceeding constructions at the surface."

Moreover, he broaches the idea of constructing undersea pressure locks that can accommodate the ingress and egress of ocean-going submarines. In his words:

> (A) submarine can sail into a water-filled lock, change the pressure, and then sail into a water compartment - into a one-atmosphere environment – in very short transit-time.

He goes on to remark that:

> Locks of this type could be used very easily on coastlines which have no harbors by drilling from shore and out under the water to a point where a cargo submarine could come in through a water-filled lock to a dock at one atmosphere, and then unload its cargo...[6]

These few brief quotations from Dr. McLean's 1969 interview, in which he ratifies the China Lake NOTS's ideas for manned, deep undersea bases accessible by submarine via deep sea locks and tunnels, suffice to convey the seriousness with which this entire

---

[5] Ibid.

[6] Ibid.

policy area was viewed at a high level within the U.S. Navy circa the mid-to-late 1960s.

With this important background context in place, let's take a look at Walter Koerschner's wonderful artwork. Keep in mind that Mr. Koerschner is an artist, not an engineer, and he was tasked with producing a body of work that would graphically depict U.S. Navy plans, as of 1968. He has told me that he has no knowledge of any facilities such as those that appear in his artwork ever having been built, and I believe him.

But that is scarcely surprising. If clandestine undersea bases do exist, then such a program would certainly be highly compartmentalized. Only personnel with a strict need-to-know would be in on it.

The Channel Islands

The Channel Islands lie just off the coast of southern California. One of them, San Clemente Island, was the site of some of the preliminary testing of the U.S. Navy's Rock-Site related experimentation. A 1967 article in a leading mining journal forthrightly displayed a photograph of a truck-mounted drilling rig at work on San Clemente Island with the following caption: "Deep coring to determine feasibility of Rock Site installation takes place on and offshore at San Clemente Island." The article was written by one of the U.S. Navy's research geologists at the China Lake Naval Ordnance Test Station.[7]

Another of the nearby Channel Islands is Santa Catalina. Interestingly, Santa Catalina Island is depicted in Mr. Koerschner's first illustration. The artist's notes say, "Illus of Catalina shore where dry installation might be placed."

---

[7] Carl F. Austin, "In the rock.... A logical approach for undersea mining of resources," *Engineering and Mining Journal*, vol. 168, no. 8 (August 1967): 82-88.

Illustration 6-1: Depiction of Santa Catalina Island Rock Site Base. *Credit:* Walter Koerschner .

Notice the small cluster of buildings just onshore, in the center-left of the frame. A shaft extends down below the buildings and extends underground and then underwater, out beneath the offshore slope of the island, as its flank submerges down into the sea. The several rooms of the in-bottom installation then open out from the shaft, with an opening at the bottom, directly to the open sea. This is a fine example of an in-bottom, manned undersea base, that has a land connection. It provides ready access to both the deep sea and to the surface, onshore environment.

How might the opening to the sea from such a facility look? In the next view, the artist shows how "high profile visitors could view underwater activity in a dry environment." Notice the access tunnel and the spherical, glass bubble viewing area. (Illustration 6-2.)

Here is another artistic representation of a "viewing capsule" for "high profile visitors". (Illustration 6-3.) Notice the hatch for access at

Illustration 6-2: Undersea capsule for high profile visitors. *Credit:* Walter Koerschner.

left, with the door lock, as well as what appears to be a vertical access shaft. Inside the glass walled viewing capsule, there is a shirt-sleeve work environment.

Larger facilities might accommodate several visitors at a time. From above, you can see what a slightly larger capsule with multiple viewing ports might look like. Notice an access hatch on top of the seafloor installation and a connecting tunnel that disappears into the side of an undersea bluff at top-center. (Illustration 6-4.)

Alternatively, a smaller spherical "viewing capsule," a sort of Sea Hut," could be constructed against an underwater bluff, with a vertical, access shaft. (Illustration 6-5.) These are some of the sorts of viewing ports that could be associated with a land-based Rock-Site underwater base as depicted in Illustration 6-1. Judging from the undersea scenery and activities shown in Illustrations 6-1 to 6-5, facilities such as these would be constructed at comparatively shallow depths, perhaps on the continental shelf, or in and on the underwater slopes of offshore islands.

Illustration 6-3: Sea Hut I. Viewing Capsule for High Profile Visitors. *Credit:* Walter Koerschner.

Illustration 6-4: Undersea Viewing Capsule for High Profile Visitors. *Credit:* Walter Koerschner.

Illustration 6-5: Sea Hut II. *Credit*: Walter Koerschner.

I cannot help but remark at this juncture that it is passing strange that my research has uncovered two explicit references showing U.S. military interest in possible undersea bases in the Channel Island region that lies just offshore of southern California. As I discuss in the next chapter, the alien abductee Linda Porter recounted a bewildering tale of being shown an undersea facility that rose up out of the seabed, somewhere offshore from Santa Barbara, California. Those who are interested may consult her account in greater detail in Linda Moulton Howe's book, *Glimpses of Other Realities, Vol. II: High Strangeness.*[8] Can it be pure coincidence that the Channel

[8] Linda Moulton Howe, *Glimpses of Other Realities, Volume II: High Strangeness*, (New Orleans, Louisiana: Paper Chase Press, 1998).

Islands lie due south and southeast of Santa Barbara? I find it significant that Ms. Porter reported being shown a strange undersea base near the very islands where the U.S. Navy showed interest in making undersea facilities. This is the sort of evidence that once again raises the question as to possible clandestine cooperation between the U.S. military and purported aliens.

Planned Undersea Submarine Bases

Returning to the wonderful illustrations by Walter Koerschner, I want to say that I absolutely love these underwater scenes. But the artwork only gets better, as in this cutaway look at a planned, deep-sea, Polaris submarine base. (Illustration 6-6.) Notice the Polaris submarine approaching the massive air lock at left. The huge submarine berth inside the sea mount is connected to crew accommodations and an under-the-seabed tunnel network by a vertical shaft and tunnel system. The first chamber near the top of the vertical shaft has two pipes that extend outside the facility to the deep sea. Based on my research I believe that such a facility would likely be powered by a compact nuclear reactor and that the pipes would be for the circulation of seawater to cool the reactor core. As you may know, at great depth seawater is very cold, and so would serve admirably to cool a nuclear reactor.

I don't know what is at the top of the vertical shaft, but it looks as if it might be a glass-domed viewing/observation/surveillance area. It would be a natural location for an ingress/egress hatch or lock, as well. The tunnel system that extends beneath the seafloor and out of the frame to the lower right is very interesting. I believe that the several vertical shafts located at the end of the short tunnels that branch off at right angles to the main tunnel are meant to be underwater nuclear missile silos. There are two reasons for my saying this. First, the 1968 SRI document that I cited earlier explicitly

Illustration 6-6: Undersea Base With Polaris Submarine. *Credit:* Walter Koerschner.

mentions that one of the purposes of deep, manned, in-bottom bases might be for "offshore missile launch sites."[9] And secondly, the U.S. Navy's Chief Scientist in the Special Projects Office from 1958-1972, Dr. John Piña Craven, wrote in his remarkable book, *The Silent War*, about Dr. Edward Teller's great interest in placing nuclear missiles in undersea glass silos at the depth of 20,000 feet. The science behind glass missile silos is fascinating. At great depth, glass has tremendous structural strength, and is a very sturdy construction material. It is counter intuitive, but nevertheless true.[10] As best as I can determine from Dr. Craven's account, his conversation with Edward Teller concerning the emplacement of nuclear missiles in all-glass silos 20,000 feet beneath the sea took place in 1964. This date is just four years prior to Walter Koerschner's

[9] Warfield and Parkinson, op. cit.

[10] John Piña Craven, *The Silent War: The Cold War Battle Beneath the Sea*, (New York: Simon and Schuster, 2002).

illustrations. Hence, I think it is a fair bet that what we are seeing here is a depiction of what Dr. Teller wanted Dr. Craven to develop and deploy on the ocean bottom. There is not the slightest shadow of a doubt, by the way, that Dr. Craven would have interacted with the manned, undersea base team at China Lake.

Other illustrations make clear that the Navy was thinking of berthing several submarines at a time within in-bottom installations. For many years, I wondered how crews could keep from going stir crazy as they silently patrolled the ocean depths for months on end – up to a half year at a time! These illustrations have gotten me to thinking about other possibilities. Who knows what submarines and their crews really do when they submerge and disappear at sea for months on end?

Illustration 6-7: Racetrack Undersea. *Credit:* Walter Koerschner.

Consider, for example, this "racetrack" facility (Illustration 6-7) – also called the "Nautilus Concept" – that can dock three submarines at a time, with an adjoining sister facility that also can handle multiple submarines. The picture is virtually self-explanatory. Large submarines are hundreds of feet long, so the dimensions of a facility such as shown here would have to be very large. The central docking area might be more than a thousand feet long and easily more than a hundred feet in diameter. The living quarters would obviously have to accommodate hundreds of crew members in some degree of creature comfort. In the bottom-center you will notice the same arrangement as in the previous illustration for circulating cooling water for a nuclear power plant. There are also two tunnels connecting the deep sea with the crew quarters. Perhaps these represent emergency ingress-egress hatches and tunnels for the living quarters.

Spiral designs also play a prominent role in the "Nautilus Concept" undersea base plans. The reason for the spiral designs is this: tunnel boring machines cannot turn a sharp corner, so when they are used to excavate a complex, all turns must be curved, the radius of the turn determined by the tunnel boring machine's

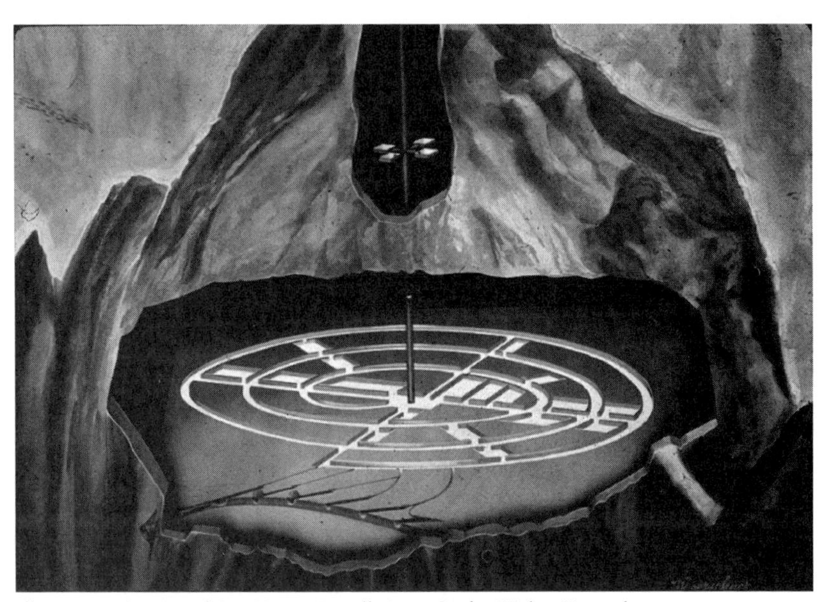

Illustration 6-8: Spiral Sea Mount Installation. *Credit:* Walter Koerschner.

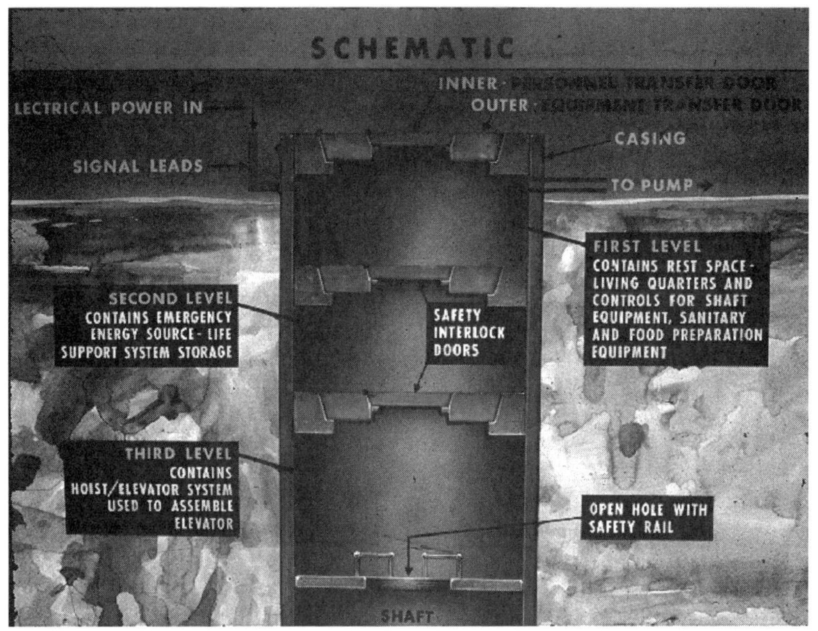

Illustration 6-9: Undersea Interlock Cutaway. *Credit:* Walter Koerschner.

turning radius. The idea would be to bring a disassembled tunnel boring machine down the central shaft shown here. (Illustration 6-8). The actual lock tube entrance to the deep shaft might have an outer lock and an inner lock, to admit both men and equipment, as in Illustration 6-9, which shows a cut-away view of a seabed airlock that grants access to the sub-seafloor environment. After all the component pieces were down the shaft, mechanics would be brought down to put the tunnel boring machine together and commence tunneling out the complex. Large tunnel boring machines can be hundreds of feet long, so the complex you are looking at could be quite large. Notice the looping entrance and exit tunnel for submarines in the lower-left-center. Excavated muck would simply be discharged outside the sea mount. (Illustration 6-8).

Layout of Other Possible Spiral-Type Facilities

Here is another possible configuration for a spiral undersea base. (Illustration 6-10.) Notice that in this case, the undersea base has an industrial function. It is an oil drilling operation, completely contained within the seabed. The rubric implies that it would be powered by a geothermal source of energy. The oil wells are arrayed in an outlying tunnel, and the petroleum they would produce is pumped up a central shaft for delivery to a waiting oil tanker on the surface of the sea, far above. I am not aware of any such operational facilities today. However, I have run across conceptual plans for such undersea operations in the mining literature that I have consulted in recent years.

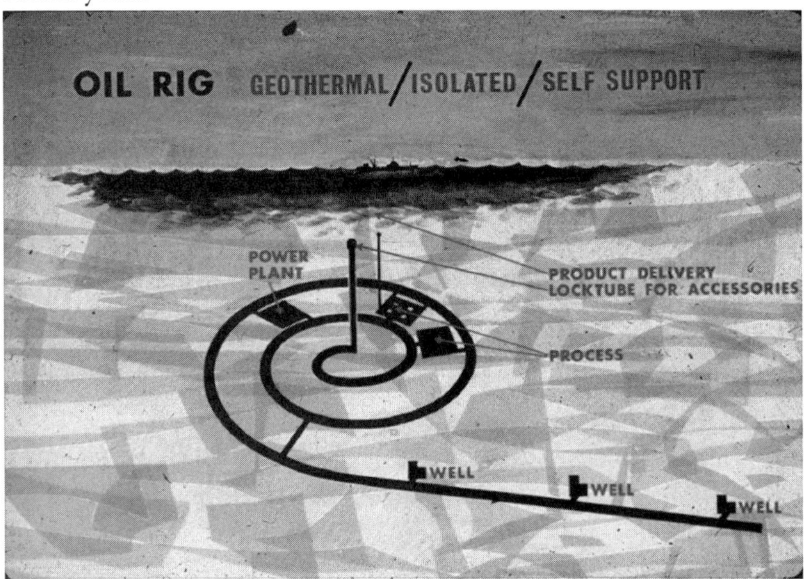

Illustration 6-10: Undersea oil rig. *Credit:* Walter Koerschner.

In yet another view of a possible, spiral-type undersea base (Illustration 6-11), you can see that there are two vertical access shafts, one in the center, and another on the left. Submarines are shown hovering around the entrance lock on the seafloor in the center, while at left another submarine departs an outlying entrance shaft. The relevant literature suggests that such a facility might contain 50 men to carry out research and active tunnel boring.[11]

The planned living and working arrangements are shown in greater detail in a separate four-part illustration. (Illustration 6-12.) Starting at upper-left, submersibles are guiding a lock-tube assembly into a finished bore-hole that has already been drilled out. The next step is to cement the lock-tube into place. Then men and equipment can enter the lock-tube and tunnel down beneath the seabed, enlarging the facility as they go. Moving clockwise, at upper-right, men are at work installing the power plant, probably a five-megawatt nuclear reactor. This is the size reactor that the U.S. Navy planned to use in its manned, in-bottom, undersea installations.[12] Continuing

Illustration 6-11: Another Spiral Sub-Bottom Installation. *Credit:* Walter Koerschner.

[11] Austin, op. cit.

[12] Ibid.

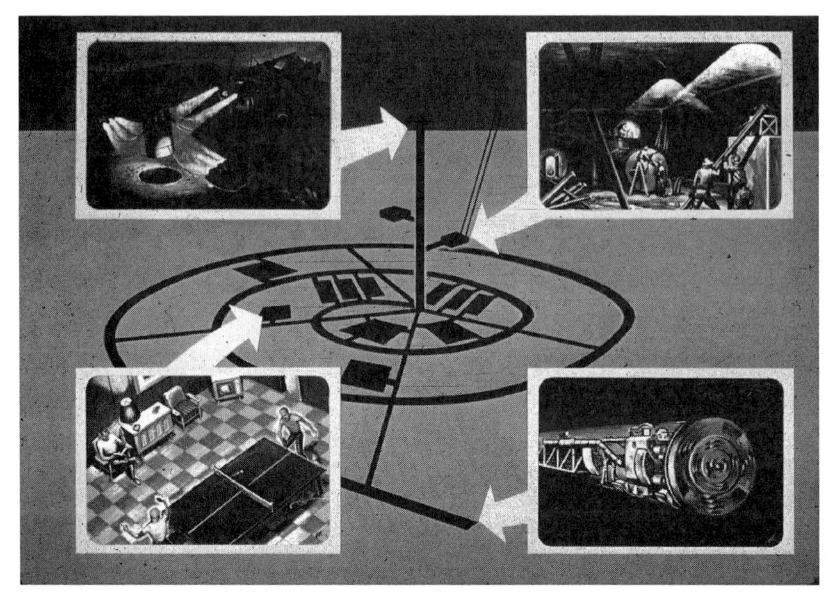

Illustration 6-12: Activities in Undersea Installation. *Credit:* Walter Koerschner.

down to the lower-right of the illustration, a tunnel boring machine is depicted grinding away, lengthening the undersea tunnel network. Finally, in the lower-left, you can see some of the station's crew playing ping-pong and reading in a recreation area.

Taking Stock

Remember, these are all artistic representations depicting the U.S. Navy's Rock-Site plans of the mid-to-late 1960s; they are not photographs of actual facilities. Neither Walter Koerschner, nor anyone else with whom I have spoken, has told me that the Rock-Site plans came to concrete fruition. And yet, we have the stories of Linda Porter and others, who tell us of mysterious undersea bases and installations that go absolutely unmentioned in the mainstream news media. What to make of this strange state of affairs?

Well, I am inclined to look at the documentation and information that my research has uncovered and extrapolate from there. It is clear that starting at least in the 1960s, the capability has existed to construct and man multiple, huge, sophisticated mid-ocean bases at great depth beneath the seafloor. This has been well discussed in the open mining and military literature, some of which is presented for your consideration in this book. But then, as we move into the 1970s, the paper trail starts to run dry and the whole subject of manned, in-bottom undersea bases appears to go away.

Except that it actually *doesn't*. As we move into the last part of the 1990s, anecdotal tales begin to pop up from people like Linda Porter and others, describing strange undersea bases that they have seen or visited. In a nutshell, it looks like once the technology was proven out, back in the 1960s, the whole thing went deep black/covert/clandestine and the real work since then has been carried out in very great secrecy. That's the way it looks to me, and if I were a betting man, I would give you excellent odds that that is exactly what has happened. What could be the reason for the extreme secrecy? I have indeed wondered whether it has something to do with a possible alien presence on this planet. Or perhaps, there are other factors just as conducive to great secrecy. I wish I could tell you I understood what is going on! But I don't. The best I can do is to present thc evidence that I have and draw probable conclusions.

The Submarine Tunnel Stories

Let's get back to Walter Koerschner's wonderful illustrations. But before I present the next couple of illustrations, I have a small confession to make. Back in the early 1990s, when I first began my underground bases and tunnels research, people began telling me that the U.S. Navy was the biggest underground excavator and tunneler of them all. These were not "insiders" from government and military programs who were telling me this, but ordinary people whom I encountered in the course of my everyday life. Somehow, I don't know how, maybe by picking up stray bits of conversation by chance, they had come to this realization that they then passed on to me. But at the time, I did not know what to make of this second- and third-hand information. So, I thought that the people telling me such uncommon things must be deluded, addled – just a few bricks shy of a full load, in other words. My incredulity only deepened when some of them told me breathlessly of alleged, secret submarine tunnels in the Long Beach, California area that purportedly ran inland from the open sea to underground submarine bases well inland. I considered such an idea almost completely beyond the pale. Absent hard proof, I was little inclined to take such stories seriously.

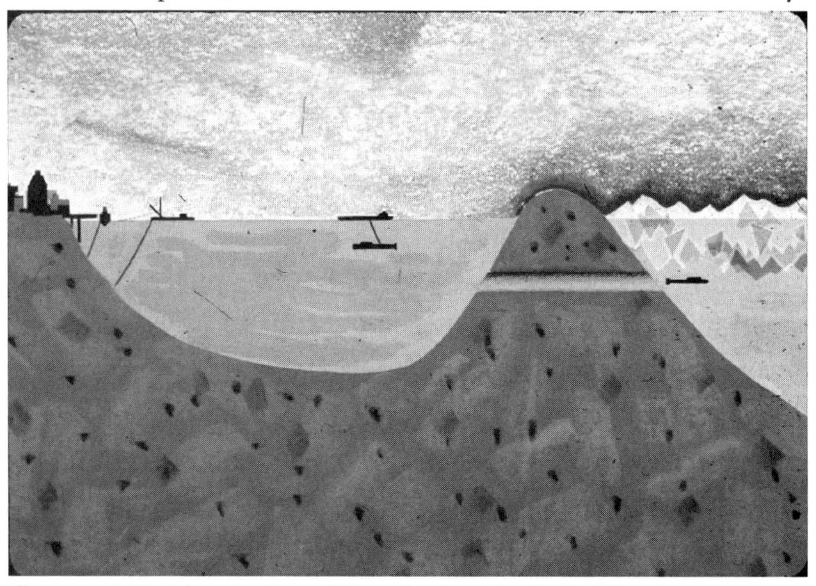

Illustration 6-13: Submarine Tunnel Through Mountain. *Credit:* Walter Koerschner.

Now, however, my mind is open a little more. Could there actually be huge, secret tunnels deep below southern California, with submarines silently gliding through them? The technology to make submarine tunnels appears to have existed for decades. So, what *if*

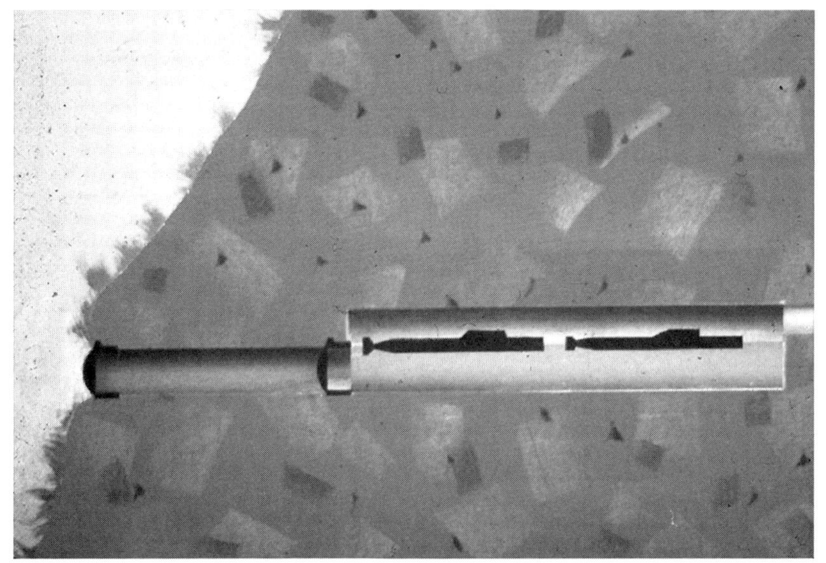

Illustration 6-14: One More Cross Section of a Sea Mountain. *Credit:* Walter Koerschner.

such structures do exist? Look at the illustration of a submarine gliding out of a tunnel that traverses a mountain, as the submarine passes from one body of water to another. (Illustration 6-13.) Presumably the scene is set in the polar regions, since the submarine is entering a sea covered by pack ice, or ice bergs. Of course, submarines have been navigating the polar regions for decades, so that is hardly a novel concept. But the idea of undersea, or underground tunnels for submarines? – ah, well, that *is* a rather novel idea that you don't encounter every day. And here it is looking you in the face in this illustration.

Take a look at the close-up of a submarine tunnel and lock system set into the side of an underwater slope. (Illustration 6-14.) I suppose in theory the submarine tunnel could be 1,000 feet long, or 20 miles long or …? Who knows if such tunnels have actually been made? If you do, please contact me with the relevant details, courtesy of my publisher.

Way Down Beneath the Moonless Mountains

Where in the world might manned, in-bottom undersea bases be found, assuming they exist? I suppose they might be just about anywhere on Earth, under just about any body of water. One such place might be beneath the sea mounts that rise from the floor of the Atlantic Ocean, to the west of Gibraltar. (Illustration 6-15.) All of them would be ideal candidate sites for construction of the type of undersea base depicted in Walter Koerschner's illustrations.

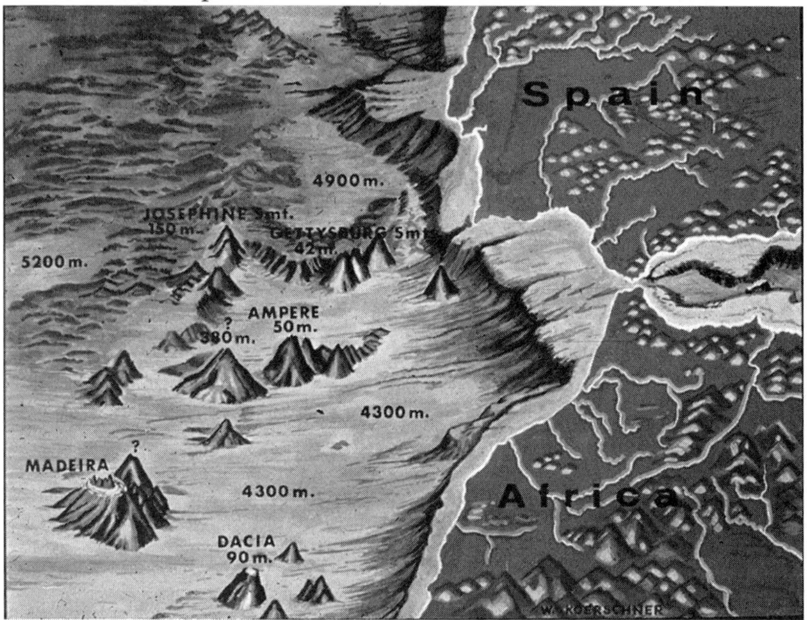

Illustration 6-15: Undersea Mounts Off of Spain. *Credit:* Walter Koerschner.

Another possible location for an undersea base might be in the Moonless Mountains, which lie deep beneath the Pacific Ocean, between California and Hawaii. (See Illustration 6-16.) The illustration shows a family paying a visit to an observation bubble on top of an undersea mount. The idea of stationing families in undersea bases was very much current in the thinking of the 1960s-era undersea base planners. I can't help but recall that the so-called

Rock-Site document that came out of the China Lake Naval Weapons Station in 1966 makes unambiguous mention of stationing crews and their families in such facilities.[13]

But why the U.S. Navy interest in the Moonless Mountains? One obvious reason might be their strategic location between the U.S. mainland and the mid-Pacific state of Hawaii. Other reasons might plausibly be scientific, such as conducting oceanographic or seismic research. In any event, I wanted to know more, so I typed the words, Moonless Mountains, into an internet search engine and quickly discovered that there is indeed an undersea observatory not far south of the Moonless Mountains. It is called the H2O Long-

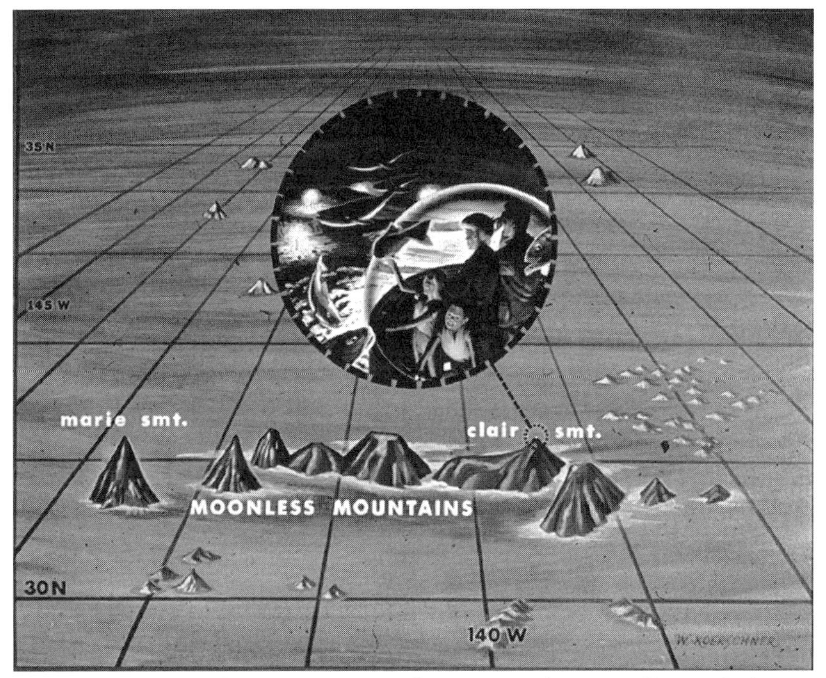

Illustration 6-16: Moonless Mountains; Family Visiting Undersea Installation. *Credit:* Walter Koerschner.

[13] C.F. Austin. *Manned Undersea Structures- The Rock-Site Concept.* NOTS TP 4162. U.S. Naval Ordnance Test Station, China Lake, California, October 1966.

Term Seafloor Observatory.[14]

I began to read and learned that there were plans to drill a borehole 325 meters deep into the seafloor and then cement a casing in place, all the way to the bottom of the bore hole. The purpose for this is to put seismic monitors into place and monitor the local geology and seismic activity. I further learned that the data from the scientific instruments will be carried back to Hawaii via the Hawaii-2 submarine cable system that was laid down by AT&T between San Luis Obispo, California and Makaha, Oahu, Hawaii in 1964. Other instruments could also be installed there to monitor geomagnetics or hydrothermal activity.[15]

This is all very fascinating and it got me to thinking what such a facility might look like, in connection with a manned, in-bottom base, especially in view of the SRI document that I cited at the beginning of this chapter. The SRI authors say forthrightly that a deep undersea, manned, in-bottom base could have "multiple uses," including military, industrial and scientific. In the appendix, the authors list one of the possible functions of a manned, in-bottom base as being "seismic monitoring." They mention this in the context of a "shallow remote" in-bottom base; however, the same technology could just as easily be deployed at a deep, remote, in-bottom base.[16]

Was a manned, in-bottom base secretly constructed under or near the Moonless Mountains, replete with submarine cables that connect it to Hawaii or California – or perhaps to other undersea bases? If it ever was, it might well resemble what you see in the next illustration. (Illustration 6-17.) Here you can see a depiction of a manned base burrowed deep down into the seabed, connected to an

---

[14] "Ocean Drilling Program, Leg 200 Scientific Prospectus, Drilling at the H2O Long-Term Seafloor Observatory," http://www.odp.tamu.edu/publications/ prosp/200_prs/200text.html, 2004.

[15] Ibid.

[16] Warfield and Parkinson, op.cit.

elaborate sonar monitoring system that is wired into a cable network running across the seafloor.

Illustration 6-17: Sonar Installation. *Credit:* Walter Koerschner.

Immediately the question arises: are there really installations like this, deep beneath the Atlantic, or the Pacific, or the Gulf of Mexico, or the Indian Ocean, or the North Sea, or the Mediterranean Sea? After slogging through many thousands of pages of pertinent documentation, I know that it is well within the state of the art of modern deep-sea engineering to make facilities such as you see here in Walter Koerschner's illustrations. The Americans can do it, the British, French, Germans and Norwegians can do it, the Russians can do it, so can the Chinese and Japanese – and if I had to guess, I would hazard to say that at least one of them, and maybe all of them and more, have done so. This does not exclude the possibility that there may also be aliens involved in similar activities, maybe on their own, or possibly in cahoots with one or more of the nationalities mentioned above. I am not saying this is what is happening, but I don't categorically rule out the possibility.

Domed Undersea Cities?

The illustration of the shark hunt (Illustration 6-18) really caught my eye. I am greatly intrigued by the glowing, translucent dome-shaped structures on the ocean floor. It is very dark at great depth in the sea, since sunlight cannot penetrate thousands of feet of seawater. Presumably, then, any structure that is sited on the deep sea bottom would need to be well lit. But the translucent, glass-like appearance of the domed structures jogged something in my memory.

A few years ago I was reading a fascinating book, *Vieques: Polígono del Tercer Tipo*, by the well-known Puerto Rican UFOlogist Jorge Martín, that had a description of a purported undersea base near Puerto Rico that was under a big, transparent dome. The story is very unusual. A man named Mr. Roldán reported being taken by

Illustration 6-18: Shark Hunt and Undersea Domes. *Credit:* Walter Koerschner.

tall, human-like alien beings to an elaborate undersea base to the east of Puerto Rico that was constructed under some kind of big "transparent cupola."[17] The book contains an illustration of this undersea city's appearance and it is most thought provoking, I assure you.

As with all stories of this type, we must consider two possibilities: a) the story is false, or b) the story may be true. More and more of these sorts of stories are popping up as the years go by, such that I am starting to suppose that perhaps some of them are true. It may even be that Mr. Roldán's story is true, although proof is surely lacking to settle the matter one way or another.

One of the things that I find interesting about his story is the reported location of this alleged undersea base east of Puerto Rico. As it happens, the U.S. Navy for many years had a large and active base on the east coast of Puerto Rico, Naval Station Roosevelt Roads (now in the process of being closed down).[18] Also, for many years the Navy conducted extensive war games on and around the nearby island of Vieques, which lies directly east of the main island of Puerto Rico. Given the decades-long history of heavy U.S. Navy activity in and around the waters that lie to the east of Puerto Rico, I seriously doubt that a deep undersea city under a transparent cupola would have gone unnoticed and unremarked upon by the U.S. military. Could such a thing possibly be true? Is it possible that such an undersea city exists, and that the U.S. Navy knows of it but says nothing? The waters around Puerto Rico are extremely deep; what lies beneath them is well beyond the limited ability of ordinary people to physically, personally investigate.

---

[17] Jorge Martín, *Vieques: Polígono del Tercer Tipo,* (San Juan, Puerto Rico: Jorge Martín, 2001).

[18] Note that the Naval Station Roosevelt Roads was slated for closure under the Fiscal 2004 Defense Appropriations Act.

That said, we do have this puzzling illustration by a U.S. Navy artist of brightly glowing domes set on the deep seabed. Walter Koerschner has clearly told me that the inspiration for these illustrations came out of his own artistic imagination. And of course that has to be true. How could it be otherwise, since he is the creative entity behind them? But at the same time, I cannot ignore that his artistic inspiration took place within the specific context of a U.S. Navy manned, in-bottom, undersea base R&D program. Is it possible that the U.S. Navy asked one of its artists to render a drawing of what a glass-domed, deep sea installation might look like, maybe even showed him some technical documents, without telling him precisely where the idea came from? Questions and more questions. We have the illustration, and we have the anecdote from Puerto Rico, and not much else. Perhaps time will reveal if there is any possible connection between the two.

Magical, Mystery Tour Concludes – Or Is it Just Starting?

So there you have it. A magical, mystery tour through the deep undersea world of Walter Koerschner. How much of what he has depicted in his illustrations has been carried out, if anything, remains unknown to me. And quite possibly to him, as well. The extremely compartmentalized world of special operations is a highly secretive, very high-tech and massively well-funded place (with untold billions of black budget dollars, of course). Perhaps the best we can hope for is to view the plans for what was on the drawing board decades ago. Maybe something like these plans has been carried out, maybe not. Maybe what has happened goes so far beyond what is depicted in these illustrations as to be like science-fiction. All I can do is present for your consideration the evidence that my efforts have so far uncovered. Make of it what you will.

Chapter 7:

And Then There Is The Alien Question...

I don't see any way around this issue: repeatedly in my underground and underwater bases and tunnels research I have encountered the theme of purported aliens. I will be completely frank with you from the outset: I positively do not know the truth behind these stories.

Having said that, this highly strange issue of purported *aliens* underground and/or underwater is exceedingly persistent. I can conceive of a whole wide gamut of possibilities, of lies within lies, conspiracies within conspiracies, insidious agendas of cover-up, concealment and deception – as well as the very real possibility that there is a substantial kernel of truth to at least some, and perhaps most, of the accounts. If the realm of clandestine underground and underwater bases and tunnels is a strange and highly secretive place (and it is!), well then, adding purported aliens to the mix and stirring up the whole kit and caboodle makes the affair even stranger!

My intention is not to draw firm conclusions on the matter, because I cannot. But what I can do, is to present for you representative information that I have encountered in the course of the last few years. What I set out for you here is not exhaustive. I have not read everything there is to read; I have not talked to everyone there is to talk to; I do not know all the stories there are to know.

But I have read *some* of what has been written in this regard, and I *have* heard some of the stories that other researchers have stumbled across.

What follows should give you plenty to think about. If even a fraction of this material is true then the universe and our place in it is far different from what our mainstream cultural and social institutions would have us to believe. And if most of it or, heaven forfend!, *all of it*, is true – then, by God, we are living in a dream world of insidiously crafted outward illusion and delusion as utterly complete in its smugly blissful, contrived ignorance as that of an illiterate, medieval peasant.

Stranger Than Dorothy's Kansas

I have read a lot of mind bending literature over the years, including my own books, which I am privileged to read before anyone else, but little of what I have read can top the first few pages of *The Ultimate Alien Agenda*, by James Walden. I picked up this intriguing little book in a used book store and have read and reread the first several pages multiple times – that's how puzzling I find the author's story.

To make a long story short, James Walden alleges that he was abducted by alien beings and taken to a secret, deep underground facility where he was subjected to an involuntary, intimate and intrusive examination procedure that was observed by upwards of 100 other beings who were in attendance. Moreover, he alleges that not all of the beings present were aliens – no indeed. Some of the other beings present were human beings, much like himself. In other words, it appeared to him that the ostensible aliens who purportedly abducted him were in cahoots with certain human beings who were in on the game – whatever this strange game may involve.

At the conclusion of the intrusive and humiliating medical examination to which he was subjected, he was informed telepathically:

> You are in an underground facility located beneath southeast Kansas.[1]

As the alien voice spoke he saw a vision of the rural Kansas countryside far above. The telepathic voice went on to say that he was "participating in a peaceful, cooperative experiment," and added that he would not be harmed. As he pondered the other humans he saw gathered around him, the telepathic voice spoke again and said:

> These human workers are volunteers who are learning to control human disease.[2]

James Walden's strange story continues on from this juncture at some length. If you want to know the rest of the details, you will have to read his book. But for my present purposes the bare bones of his story as presented above are quite fascinating enough.

But can it be true? Are there really secret underground bases deep beneath the American Midwest in which aliens and human researchers work in great secrecy, cheek by jowl, and to which unsuspecting humans such as James Walden are abducted, there to be poked and prodded – as if they were nothing more than laboratory subjects in a medical school amphitheater?

I don't know the answer to this question. However, more and more people seem to be coming forward to report this sort of experience as the years pass. It certainly appears possible that highly covert and uncommonly strange goings-on might be transpiring

---

[1] James L. Walden, *The Ultimate Alien Agenda: The Re-engineering of Humankind* (St. Paul, Minnesota: Llewellyn Publications, 1998).

[2] Ibid., p. 7.

underground. I am reminded, in this regard, of the research done by the Austrian husband-and-wife research team of Helmut and Marion Lammer, and their book, *MILABS*.[3]

Joint Military and Alien Abductions to Underground Bases?

In *MILABS*, the Lammers examine the many arcane aspects of the American military's research into mind control technology and experimentation. Helmut and Marion Lammer point out that a number of people who reported being abducted and taken to clandestine underground bases claimed to see both human military personnel and aliens present together. Others reported being taken underground by means of technology that seemed alien, where they encountered yet more alien-seeming technology. The reasons for these experiences and the possible reality behind these unusual reports remain perfectly obscure. Whatever is going on, it is largely kept hidden from public view; many of the stories only filter out at the periphery of consensual reality, in books such as *MILABS*, and *The Ultimate Alien Agenda*, that are issued by small, unconventional publishers who are willing to take a chance on subject matter that the huge publishing houses in New York and London won't touch.

Here is just one of the thought provoking stories that the Lammers include in *MILABS*. The account stems from the experiences of a pseudonymous woman named Evelyn.

> ... I can remember flying in a beautiful, golden ship and it went into a mountainside and flew under the ground, into a huge cavernous room, where to my surprise, were human beings, dressed in military uniforms, with machine guns slung behind their backs. I walked in a line with some other people and we passed through an "energized" gate that somehow does

[3] Helmut and Marion Lammer, *MILABS!: Military Mind Control and Alien Abduction* (Lilburn, Georgia: IllumiNet Press, 1999.)

> something to your atoms. I remarked to myself: "The military have some alien technology.[4]

This is certainly a most unusual story. At first blush it seems like such a radical departure from consensual reality that it is not credible, and is more likely a fantasy or a surreal dream of some sort. But what if it is true? What if this brief account as told by "Evelyn" faithfully reflects an experience that really happened to her? I pose this question, because in recent years I have heard and read more unusual stories like this than I would have formerly dared think possible. I can see two possibilities: either an awful lot of people are telling wonderfully entertaining and confabulated fibs, or something very strange is going on involving *aliens* (whoever and whatever they may ultimately prove to be). Moreover, aliens that may be working with covert elements of the United States military, underground and possibly undersea.

Alien Bodies Kept Underground in California?

In the fascinating book, *Glimpses of Other Realities, Volume II: High Strangeness*,[5] independent journalist and author Linda Moulton Howe cites the puzzling experiences of an abductee named Linda Porter. I want to mention two aspects of Linda Porter's story that are directly germane to the underground and underwater bases theme of this book. At the same time, I also want to say that Linda Porter's reported experiences resonate to a degree with certain aspects of my own research, such that I am inclined to listen to what she has to say with a little more care than I otherwise might. Ms. Porter's experiences have to do in part with an alleged secret underground base in

[4] Ibid., 88-89.

[5] Linda Moulton Howe, *Glimpses of Other Realities, Volume II: High Strangeness* (New Orleans, Louisiana: Paper Chase Press, 1998). Although another edition of *Glimpses of Other Realities* was subsequently published in 2001 by LMH Productions as ISBN 0-9620570-3-7, I cite only the Paper Chase Press edition here.

the San Diego, California area and another alleged base underwater, on the seabed off the California coast. I find this fascinating, because based on my own evidence and informed hunches it would be my educated guess that the United States Navy, as well as other agencies and organizations, whether publicly known or completely unknown to the public-at-large, have secretly constructed and operate underwater bases in California coastal waters. I also surmise that there are underground bases onshore, not only in the San Diego area, but elsewhere in California, for example at the massive U.S. Navy installation at China Lake, near the town of Ridgecrest.

With regard to the underground base near San Diego, Linda Porter alleges that her alien abductors told her that there was an underground facility called "the Sycamore Remote Facility run by General Dynamics." In this extremely secure underground facility, the United States government allegedly held alien beings in suspended animation, in transparent computer-monitored containers. One particular human-like being held in this facility was of special interest to the aliens who wanted him back. Adding further to the air of mystery surrounding this story, Linda Porter says that there are:

> ... underground tram systems that lead from the naval base in San Diego to this facility and to another place.[6]

This "other place," according to Porter, is a normal looking house with an attached garage that is the actual entrance to the underground facility where the bodies of the alien beings were allegedly being held. If her story is true, then this would be very clever concealment, indeed.

Linda Porter also describes an undersea facility. Her description positively fascinates me:

---

[6] Ibid., 249-250.

> I was supposedly taken to an underground base off the coast of California. For some reason I was led to believe it was in the Santa Barbara area. If you were standing on sand at the bottom of the ocean (where this place is), all you could see is what looks like a silver submarine conning tower rising up out of the sand. And it would probably be the height of a two or three story building. I was told this tower thing was camouflaged by an electronic net of some kind that renders it invisible. And they also have something around it that seems to repel people and fish for some reason.
>
> Inside the building the floors and the walls and the ceiling were all a silver-grey color. There was a lot of light. But the doors – and there seemed to be doors all over the place – were brightly colored. They were either bright red or bright blue or bright yellow. And over each door was some kind of writing that looked like hieroglyphic or Arabic writing or something like that.[7]

No doubt there will be those who will read Linda Porter's stories about her reported experiences and dismiss them as too fantastic to be credible. But I must say that I am not among that number. Indeed, I believe that Linda Porter's stories may very well contain a solid kernel of truth. Here is why I find it easy to believe that the U.S. Navy might have a clandestine underground facility near its important base at San Diego, and why there may even be multiple underground bases in that region of California. Simply put, my research reveals unambiguously, via the U.S. Navy's *own* documentation, that the Navy has an interest in constructing underground facilities, and that the interest goes back many years into the past. Given the Navy's long-time presence in the San Diego area, and its technical capabilities, it should be veritable child's play to construct such facilities in and around San Diego.

Similarly, when I read Ms. Porter's account of the alleged undersea base off the coast of California, I instantly recognized that what she describes is perfectly consonant with the technology for constructing a manned undersea base that was put forth in the U.S.

---

[7] Ibid., 244.

Navy "Rock-Site" document from 1966.[8] In that document, the Navy proposed building undersea bases by cementing a huge, metal cylinder or pipe into the ocean floor, and then using it as an entrance to the sea bed. From this, a large, manned, undersea base could be constructed.

In other words, Ms. Porter describes seeing an underwater base off the coast of California that corresponds *almost exactly* to the planned entrance to a United States military "Rock-Site" underwater base – as described decades earlier in an official, U.S. Navy document. I find this detail significant.

What if the U.S. Navy has secret underwater and underground bases where strange activities transpire? Activities that may strain the boundaries of what the "dumbed-down" mass mind of modern popular culture considers to be possible? At this point in my research, I am open to that possibility. I am willing to hear what Linda Porter has to say, to file the data, to compare and contrast what she reports with the information that my own research has uncovered. And I am willing to entertain the idea that the world just may be a far stranger place than the majority of people are prepared to believe.

Underground Alien Communication Training?

Other possible evidence for covert dealings between the United States military and purported alien beings (whoever or whatever they may finally prove to be, and from wherever or whenever they may hail) appears in an enigmatic little book by Dan Sherman. The book is entitled *Above Black: Project Preserve Destiny*,[9] and purports to be a true-life account of Mr. Sherman's experiences. In a nutshell, Dan Sherman is an ex-Air Force sergeant who maintains that he was

[8] Author's note: In addition to discussing this document in this book, I cited it at length in my 2001 book, *Underwater and Underground Bases*.

[9] Dan Sherman, *Above Black: Project Preserve Destiny* (Wilsonville, Oregon: OneTeam Publishing, 2001).

covertly trained in a clandestine underground facility in Maryland as an intuitive communicator between the United States National Security Agency (NSA) and alien beings. He served in the Air Force as an electronic intelligence specialist and says that he performed duties for the NSA on the side as an alien communicator, using his regular Air Force duties as a cover.

His saga began when he was ordered to Fort Meade, Maryland, where the NSA is headquartered, to attend an electronic intelligence course. As it turned out, he attended the assigned course and much *more.* In the evenings, after his regular classes had finished, he was driven to an unknown underground facility which he entered via an elevator ride straight down. There, he was trained to mentally communicate with aliens. After a three-year period as an alien communicator, he began to receive communications that he suspected to have something to do with the abduction of humans by aliens. This suspicion gnawed at him. Ultimately, he decided to abandon his participation in the alien communication project and to leave the Air Force.

Dan Sherman's is yet another strange story, a slightly different permutation of the increasingly familiar elements of extreme military secrecy, alien beings, and a clandestine underground facility. Story after story, book after book, the personal experiences keep coming.

What on Earth is going on? Or, should I say, what *under* Earth is going on?

What About Area 51?

Unsubstantiated stories about alleged aliens at Area 51 have been making the rounds of the UFOlogical rumor mill for years. I have heard and read stories that purport to tell of aliens held captive there, or of exotic flying saucers stored there, undergoing back-engineering and flight testing. The truth of these sorts of stories remains very much in question. And still they keep coming.

One of the most interesting of these accounts that I have seen appeared in an interview in *Nexus Magazine*.[10] David Adair, a former teenage engineering prodigy, alleges that he designed a type of fusion rocket prototype that came to the U.S. Air Force's attention back in early 1971. The Air Force was so interested in what he had done, that they bundled him off to Area 51 for a personal consultation. Once at Area 51 they took him underground to examine what he surmised was a real fusion rocket engine; albeit one that he believes must have been alien technology, owing to its large size and sophisticated, exotic engineering.

The elevator that took him and his escorts underground was unbelievably mammoth – the size of a football field. It was clearly designed to raise and lower very large and heavy equipment. According to Adair, the elevator platform was supported by a dozen or more giant worm screws, the physical dimensions of which he compared to sequoia trees. These enormous worm screws could bear an extremely heavy load and slowly dropped the entire platform straight down about 200 feet to a vast underground work area. The subterranean work area was so stupendously large that it seemed to him that it could accommodate one hundred Boeing 747 jumbo jets with plenty of room to spare. The underground room had a huge ceiling and stretched completely out of sight into the distance. On the way to examine the alien fusion engine he was escorted past a series of exotic aircraft, some of which he recognized, others of which he did not.

---

[10] Robert M. Stanley, "Electromagnetic Fusion and ET Space Technology," *NEXUS New Times Magazine* USA/Canadian Edition, vol. 9, no. 5 (September-October 2002): 53-57,74-75.

Adair alleges that once he examined the novel fusion engine, which was the size of a bus, he came to the following conclusions: 1) it was of extraterrestrial manufacture; 2) it had to have been built in deep, intergalactic space; 3) it was a self aware, living machine that seemed to be an amalgam of organic and inorganic components; 4) its manufacture was seamless, as though it had been grown, not built; and 5) the U.S. Air Force appeared to not understand the engine.

And to top it all off, at the time he was only seventeen years old, and therefore not of majority age. This matters, because as a minor he could not legally be held to a National Security Oath and thereby prevented from talking about what he saw and experienced underground at Area 51 in 1971.

This is quite a dramatic story that Adair tells. I personally communicated with Robert Stanley, who conducted the interview, and Mr. Stanley indicated to me that he believes David Adair is telling the truth. And perhaps he is. Based on my documentary and archival research and conversations with a variety of sources, it is a 100% certainty that there are huge underground workings beneath southern Nevada, including in and around Area 51. So the mind boggling size of the secret, underground work area that David Adair describes presents nothing that contradicts my research. As to his even more spectacular allegation of ostensible alien technology of unknown origin secretly sequestered underground—that is a more difficult issue for which I do not have a ready answer. Nevertheless, I have included David Adair's story because I believe it just may be true.

I would hasten to add that if Adair's story *is* true, and if the other stories of aliens or alien technology underground are equally true, then one very powerful motive for the high degree of secrecy that conceals clandestine underground and undersea activities becomes quite clear: to cover up extraterrestrial and/or alien realities. The prime impetus to secrecy then becomes social control, and not national security. It's hard to control people's minds and thoughts (and hence their bodies and actions) once they firmly grasp that not only are they not alone in the universe, they are not even alone on their home planet. And perhaps never have been. What if that is one of the big secrets buried underground and underwater?

Unearthly Disclosure

The best-selling British author and researcher, Timothy Good, has written a series of fascinating books dealing with the question of a possible alien presence on Earth. I won't take the trouble of summarizing the entire body of his work; suffice it to say that Good has literally traveled the world in pursuit of answers to this intriguing question. He has spoken to myriad knowledgeable individuals, in and out of government, and come to a carefully considered conclusion.

To wit, Good has come down in support of a probable alien presence on Earth, albeit a rather crafty presence. At the end of one of his books, *Unearthly Disclosure,* he states:

> I, for one, would welcome an official disclosure to the effect that we share this planet with denizens of other planets.[11]

Earlier in *Unearthly Disclosure,* Good devotes several passages to information he has received from a variety of private individuals and government sources -- information having to do specifically with purported alien bases said to be located both underground and undersea. He discusses a variety of locations around the world for these alleged undersea and underground alien bases, including the Commonwealth of Puerto Rico and adjoining waters of the Caribbean Sea, various locations beneath the Pacific Ocean, as well as a

---

[11] Timothy Good, *Unearthly Disclosure* (London, UK: Arrow Books, 2001), p. 323.

number of states in the USA, including New Mexico, West Virginia and Alaska, and also at Pine Gap, Australia.

Timothy Good is a careful researcher; his writing is deliberate and measured. It is my view that his evidence and conclusions lend credence to the idea that there is an alien presence on Earth, and that clandestine underground and undersea facilities are part and parcel of that presence. Moreover, in mentioning the vicinity of the American military intelligence base at Pine Gap as an alien underground base site,[12] we see once again the idea of American military knowledge of an alien presence, if not interaction between the American military and certain alien factions.

This is a theme that I have encountered repeatedly in my research. It may be a disinformation ploy intended to deflect serious scrutiny of clandestine activities that various agencies and organizations wish to conceal from public view. Alternatively, maybe there truly are aliens on Earth and maybe the American military really does have knowledge of them and really does secretly interact with them. A growing body of evidence suggests this is the case.

Is This The Real Flying Saucer?

In June of 1955 an extraordinarily thought provoking article appeared in *Look* magazine. I was born in 1955 and remember very well the American Golden Era of the 1950s and early 1960s – when great change was in the air. The atmosphere seemed charged with an expectant air of imminent breakthrough and unbridled opportunity of every conceivable description. It was an exciting period of dynamic social ferment and rapid technological and scientific advances. And then John Kennedy was assassinated by an evil cabal at the heart of the U.S. power structure, and the American dream guttered out irretrievably into a thousand dying embers. Something happened

---

[12] Ibid., 315-316.

that year to poison the shining promise that was the America of yesteryear. The result, sadly, was that the America of my adolescence and adult years was stillborn. The glorious America that could have been was dead on arrival. Many years later, America and the rest of the planet are still suffering the catastrophic consequences, as we spiral ever downward into a vicious global cycle of never-ending military attack and counterattack.

Is it possible that one of the great advances of that period involved the construction of vast, secret underground bases that housed fleets of flying saucers? Could that have been the "something" that strangled the American dream in its cradle? Is that one of the great secrets that the covert underground and underwater bases are hiding? Have huge advances and discoveries really been made – and kept hidden in enormous bases deep down beneath the surface of the land and sea, away from the open gaze of the public who have been fooled into silence and ignorance?

This question must be asked because of the information presented in *Look* magazine in 1955. Beginning in the immediate post-WWII years with the Kenneth Arnold UFO sighting and the rumors of a crashed flying saucer near Roswell, there has been an unending stream of unexplained UFO sightings and encounters worldwide that has continued through the present day. The 1950s were certainly no exception. Though I was just a small child, I can remember very well that UFOs were seen in Tidewater Virginia, near my boyhood home.

So it was not unusual for a popular magazine to ask, as *Look* did in 1955, "Is This The Real Flying Saucer?"[13] Many people were curious about flying saucers in those days, and the magazine was

[13] Ben Kocivar, "Is This The Real Flying Saucer?", *LOOK* vol. 19, no. 12 (14 June 1955): 44-46.

simply presenting material to its readership that its editors knew they wanted to read.

But, oh my goodness, was it ever an interesting article! To read it more than half a century later is to ponder what might have been made public – but perhaps has been hushed up instead. Look at the illustration.

Where in the world did the idea for a huge, underground flying saucer base (camouflaged beneath a rugged mountain chain, no less) come from? The idea of secret underground bases did not have popular currency in 1955 the way it does now. And yet, look at the illustration. Where did the idea come from?

In the first paragraph of the article, the author states that no flying saucer has been captured and no government has taken credit for building them. But then in the second paragraph, he gets right down to brass tacks and lets fly with this broadside: "...persistent and fairly credible rumors recur that a Canadian aircraft manufacturer, A.V. Roe, Canada, Ltd., has had a saucer design under development for two years."[14] And the article continues: "The A.V. Roe people maintain a confusing silence about the whole thing."[15]

Most curious. This same company has surfaced elsewhere in my research. During the 1970s, it owned the abandoned, undersea Dosco mine in the Canadian Maritime provinces. This mine was explicitly mentioned in a 1960s-era report, cited in my book *Underwater and Underground Bases,* as a potential candidate for a prototype manned, U.S. Navy sub-seafloor base. How interesting that in the previous decade, the same company was mentioned in a popular article that discussed housing fleets of flying saucers in large underground bases with camouflaged take-off shafts. Is this just a bizarre coincidence, or is there a connection? Is it possible that the

---

[14] Ibid., 44.

[15] Ibid., 44-45.

Illustration 7-1: The original *LOOK* magazine article supplied the following caption for this illustration: "Future airports built for vertically rising saucers would have no need of the long, vulnerable runways today's fighters require. The complete operation could go underground. Tunnels with take-off shafts set into the ground, complete with maintenance bays, fuel and crew quarters, would be bombproof shelters for a saucer squadron. The shafts would be sealed after take-off for camouflage and protection." *Source*: *LOOK*, 14 June 1955.

A.V. Roe Company actually did develop a working flying saucer technology, and that they did base such machines in clandestine underground or undersea bases? The documentary evidence trail is curious enough that the question begs to be asked.

The *Look* article also commented on the desirability of having a fast fighter aircraft that could maneuver easily and *vertically*, even at high altitudes. It even provided a large and detailed sketch of what such a craft might look like, and closes with these enigmatic words:

> ...based on the current requirements of our defense effort and the demonstrated abilities of our engineers, an educated guess is that a flying saucer much like this one may well be flying within the next few years.[16]

How *very* interesting. But suffice it to say that such a flying saucer, if actually built, has been deployed in great secrecy, because nothing like that has ever been publicly displayed. Quite to the contrary, A.V. Roe made a very splashy, public display at the time of its *failure* to

---

[16] Ibid., 46.

produce a working flying saucer. The whole A.V. Roe flying saucer story quickly dried up and went silently away.

Interestingly, people the world over have continued to see many UFOs, including flying saucers, through the years to the present day. Moreover, my research has conclusively demonstrated the concrete existence of covert underground bases, and the highly probable existence of clandestine, manned, sub-seafloor bases. Could it be that A.V. Roe really did make a working flying saucer back in the 1950s, and that the technology has been kept secret, for covert use only? Is it possible that some of the flying saucers seen by people are actually made with great stealth by the terrestrial, human military-industrial complex? And might some of them be based in great secrecy in concealed underground and undersea bases? I wonder, I really do.

The question remains, over half a century later: what did the author know about flying saucers and underground bases, and what led a prominent magazine like *Look* to publish such a feature article? And where did these enigmatic ideas come from in the first place? It is all very mysterious.

Is Something Hidden Deep Beneath The North Sea?

Needless to say, the plot only thickens with reports of unconventional craft seen entering and leaving the waters of the North Sea, between the British Isles and the European mainland.[17] The geology beneath the North Sea is certainly conducive to the construction of undersea facilities and tunnels. Depending on location, there is a solid stratum of chalk that lies approximately 2,000 to 3,000 feet beneath the sea floor. This thick chalk bed underlies most of the North Sea. As it happens, this is a favorable environment for the operation of mechanical tunnel boring machines. No doubt, many readers will recall that when the Chunnel was bored between France

[17] Private communication with Graham Birdsall, editor of *UFO Magazine* (U.K.).

and England, that the tunneling machines were digging for miles through a sub-sea chalk deposit that lies beneath the Channel. This is not surprising, given that the English Channel is the southernmost extension of the North Sea. The sub-sea chalk bed is widespread in the region and would present the same favorable environment for tunnel boring elsewhere that it does between England and France.

The question then becomes whether other sub-sea tunnels have been covertly bored beneath the North Sea. I must say that I believe the possibility is very real. Indeed, it may be that the unconventional flying objects seen coming and going from the North Sea are traveling to and from hidden, sub-sea bases and tunnels. I think that serious investigators must keep this possibility in mind.

Conclusion

By now, I am clear on a few simple facts. 1) There are many secret underground and probable undersea bases and tunnels. 2) They can be impressively large and surprisingly deep. 3) They contain very sophisticated technology. 4) Many thousands of people are involved in constructing, operating, and maintaining these facilities. 5) They are located all over the world. 6) Aliens are probably involved in some, or maybe many, of the underground and undersea bases. 7) High-speed tube shuttle train systems probably do exist. 8) We, the people of this planet, are being massively lied to by our so-called "leaders".

One of the best known underground facilities in the world is at Pine Gap, near Alice Springs, in the geographical middle of the vast Australian outback. Even though the base is in Australia, it is managed and run by the American military and espionage "alphabet soup" agencies. Officially, the Pine Gap facility is a spy satellite intelligence facility used by the Americans to monitor global satellite communications. If you type the key words "Pine Gap" or "Pine Gap Base" in a major Internet search engine such as Google or Yahoo, you will get thousands of responsive links. Not all of these links have to do with the base at Pine Gap, Australia, but many of them do. You could spend years investigating just this one base. However, short of joining the CIA and signing your life and soul away to enter the black world of compartmentalized, special operations, you would never be permitted to enter the underground levels at Pine Gap.

That having been said, for 20 years, maybe more, I have been hearing rumors and stories about a large underground base at Pine Gap. I find the stories perfectly credible, because the U.S. military and espionage agencies have a well documented history of creating underground bases in the United States. There is no reason why they wouldn't do abroad what they do in the USA; on the contrary, there is every reason to think that they would. One of the most eye-popping of these stories was related to me several years ago, second-hand. It was so outlandish that at first I thought it was so far beyond the pale that it could not possibly be true. But I have to confess that as my knowledge of the missing trillions of dollars from the coffers of the Pentagon and other American government agencies has grown, and as I have come to appreciate the mafia-like nature of the U.S. Federal government and the mega-corporations and multinational conglomerates with which it transacts all manner of undertakings, from A to Z, my mental horizons have broadened to encompass thoughts and levels of understanding to which I was simply unconscious in the past. According to my source, sometime around 1981 the base at Pine Gap was planned to be extended vertically to a depth of 8 miles, with 400 levels, or "floors" if you will, spreading out laterally underground across 20 or 25 miles, and designed to ultimately accommodate 250,00 inhabitants or more.

In other words, the allegation is that the Shadow Government intended to make a mid-size city underground at Pine Gap. I now find myself wondering if they did make an underground city there. And if the answer is in the affirmative, who lives there? Do they live there voluntarily of their own free will, or are people held there against their will in captivity? And what is the purpose of such a large facility, assuming such a big, underground city has been constructed?

The United States government has lied to its citizens and the world about so many policies – so often, so massively, for so many years – and I find myself in possession of such a wealth of persuasive evidence of clandestine underground facilities and undersea bases, that I can no longer put any arbitrary, preconceived limits that I might have once held on the true extent and size of the secret underground and undersea bases and tunnels on this planet. Secrecy and compartmentalization have run completely amok, funded with untold sums of black budget money, and enabled by jaw dropping, cutting edge technology straight out of a science fiction movie.

We, the so-called ordinary people of this world, are faced with a simple issue: we need to reclaim ownership of our planet and of our own lives from these darkly sinister, deeply wicked Under Lords of Planet Earth. They are stealing our heritage, ruining our planet, plunging us into life-long debt peonage, and imposing a global, Orwellian, totalitarian, social control structure designed to enslave as many human beings as possible, body, mind and soul, in a cradle to grave, open air prison. It was almost 25 years ago that I first encountered the idea of "Prison Planet Earth" – but I now see that that is what is being very rapidly set up. The intent of these demonic powers is to turn the entire planet into a giant prison, and to enslave all of us under their harsh dominion, for all time to come, and to do this to us on the planet of our birth.

So our moment of truth has arrived – the years 2010 to 2012 are a make or break watershed period for us. What will our response be? Do we continue to passively acquiesce in the twisted pathology of the global Shadow Government, and remain silent as we are stripped of our freedoms and plunged into Big Brother slavery, or do we rise together and say: "Enough! We're not going to take any more of your lies, thievery, criminal corruption and abuse! We're reclaiming control of our own lives and of the planet we inhabit!"

It's that simple. That's our choice.

Index

About the Author:

Richard Sauder has been researching the mysterious topic of secret underground and underwater bases and tunnels since 1992. He is the author of four previous books on these and other topics. Dr. Sauder holds degrees in sociology, political science, Latin American studies, and forestry. He is a native Virginian who has lived in many American States within the Mid-Atlantic region, the South, and the Southwest.

ANTARCTICA AND THE SECRET SPACE PROGRAM
By David Hatcher Childress

David Childress, popular author and star of the History Channel's show *Ancient Aliens*, brings us the incredible tale of Nazi submarines and secret weapons in Antarctica and elsewhere. He then examines Operation High-Jump with Admiral Richard Byrd in 1947 and the battle that he apparently had in Antarctica with flying saucers. Through "Operation Paperclip," the Nazis infiltrated aerospace companies, banking, media, and the US government, including NASA and the CIA after WWII. Does the US Navy have a secret space program that includes huge ships and hundreds of astronauts?

392 Pages. 6x9 Paperback. Illustrated. $22.00 Code: ASSP

NORTH CAUCASUS DOLMENS
By Boris Loza, Ph.D.

Join Boris Loza as he travels to his ancestral homeland to uncover and explore dolmens firsthand. Chapters include: Ancient Mystic Megaliths; Who Built the Dolmens?; Why the Dolmens were Built; Asian Connection; Indian Connection; Greek Connection; Olmec and Maya Connection; Sun Worshippers; Dolmens and Archeoastronomy; Location of Dolmen Quarries; Hidden Power of Dolmens; and much more! Tons of Illustrations! A fascinating book of little-seen megaliths. Color section.

252 Pages. 5x9 Paperback. Illustrated. $24.00. Code NCD

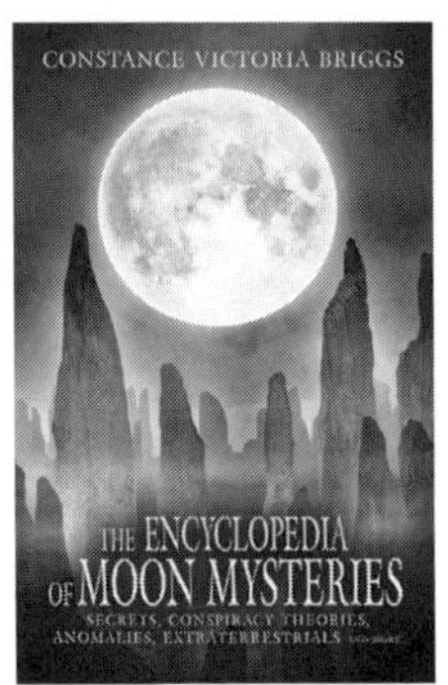

THE ENCYCLOPEDIA OF MOON MYSTERIES
Secrets, Anomalies, Extraterrestrials and More
By Constance Victoria Briggs

Our moon is an enigma. The ancients viewed it as a light to guide them in the darkness, and a god to be worshipped. Did you know that: Aristotle and Plato wrote about a time when there was no Moon? Several of the NASA astronauts reported seeing UFOs while traveling to the Moon?; the Moon might be hollow?; Apollo 10 astronauts heard strange "space music" when traveling on the far side of the Moon?; strange and unexplained lights have been seen on the Moon for centuries?; there are said to be ruins of structures on the Moon?; there is an ancient tale that suggests that the first human was created on the Moon?; Tons more. Tons of illustrations with A to Z sections for easy reference and reading.

152 Pages. 7x10 Paperback. Illustrated. $19.95. Code: EOMM

OBELISKS: TOWERS OF POWER
The Mysterious Purpose of Obelisks
By David Hatcher Childress

Some obelisks weigh over 500 tons and are massive blocks of polished granite that would be extremely difficult to quarry and erect even with modern equipment. Why did ancient civilizations in Egypt, Ethiopia and elsewhere undertake the massive enterprise it would have been to erect a single obelisk, much less dozens of them? Were they energy towers that could receive or transmit energy? With discussions on Tesla's wireless power, and the use of obelisks as gigantic acupuncture needles for earth, Chapters include: Megaliths Around the World and their Purpose; The Crystal Towers of Egypt; The Obelisks of Ethiopia; Obelisks in Europe and Asia; Mysterious Obelisks in the Americas; The Terrible Crystal Towers of Atlantis; Tesla's Wireless Power Distribution System; Obelisks on the Moon; more. 8-page color section.

336 Pages. 6x9 Paperback. Illustrated. $22.00 Code: OBK

BIGFOOT NATION
A History of Sasquatch in North America
By David Hatcher Childress

Childress takes a deep look at Bigfoot Nation—the real world of bigfoot around us in the United States and Canada. Whether real or imagined, that bigfoot has made his way into the American psyche cannot be denied. He appears in television commercials, movies, and on roadside billboards. Bigfoot is everywhere, with actors portraying him in variously believable performances and it has become the popular notion that bigfoot is both dangerous and horny. Indeed, bigfoot is out there stalking lovers' lanes and is even more lonely than those frightened teenagers that he sometimes interrupts. Bigfoot, tall and strong as he is, makes a poor leading man in the movies with his awkward personality and typically anti-social behavior. Includes 16-pages of color photos that document Bigfoot Nation!

320 Pages. 6x9 Paperback. Illustrated. $22.00. Code: BGN

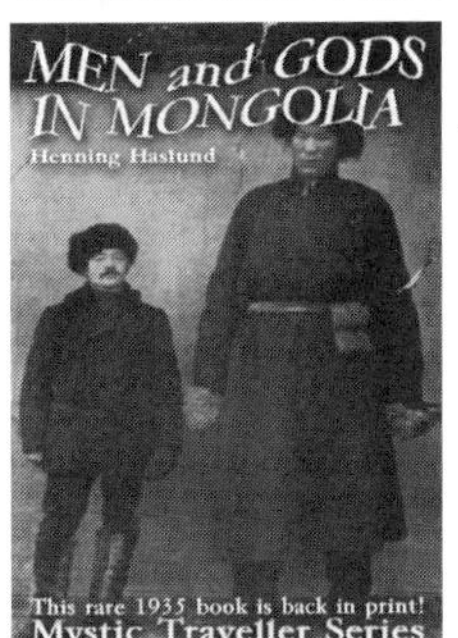

MEN & GODS IN MONGOLIA
by Henning Haslund

Haslund takes us to the lost city of Karakota in the Gobi desert. We meet the Bodgo Gegen, a god-king in Mongolia similar to the Dalai Lama of Tibet. We meet Dambin Jansang, the dreaded warlord of the "Black Gobi." Haslund and companions journey across the Gobi desert by camel caravan; are kidnapped and held for ransom; withness initiation into Shamanic societies; meet reincarnated warlords; and experience the violent birth of "modern" Mongolia.

358 Pages. 6x9 Paperback. Illustrated. $18.95. Code: MGM

PROJECT MK-ULTRA
AND MIND CONTROL TECHNOLOGY
By Axel Balthazar

This book is a compilation of the government's documentation on MK-Ultra, the CIA's mind control experimentation on unwitting human subjects, as well as over 150 patents pertaining to artificial telepathy (voice-to-skull technology), behavior modification through radio frequencies, directed energy weapons, electronic monitoring, implantable nanotechnology, brain wave manipulation, nervous system manipulation, neuroweapons, psychological warfare, satellite terrorism, subliminal messaging, and more. A must-have reference guide for targeted individuals and anyone interested in the subject of mind control technology.

384 pages. 7x10 Paperback. Illustrated. $19.95. Code: PMK

LIQUID CONSPIRACY 2:
The CIA, MI6 & Big Pharma's War on Psychedelics
By Xaviant Haze

Underground author Xaviant Haze looks into the CIA and its use of LSD as a mind control drug; at one point every CIA officer had to take the drug and endure mind control tests and interrogations to see if the drug worked as a "truth serum." Chapters include: The Pioneers of Psychedelia; The United Kingdom Mellows Out: The MI5, MDMA and LSD; Taking it to the Streets: LSD becomes Acid; Great Works of Art Inspired and Influenced by Acid; Scapolamine: The CIA's Ultimate Truth Serum; Mind Control, the Death of Music and the Meltdown of the Masses; Big Pharma's War on Psychedelics; The Healing Powers of Psychedelic Medicine; tons more.

240 pages. 6x9 Paperback. Illustrated. $19.95. Code: LQC2

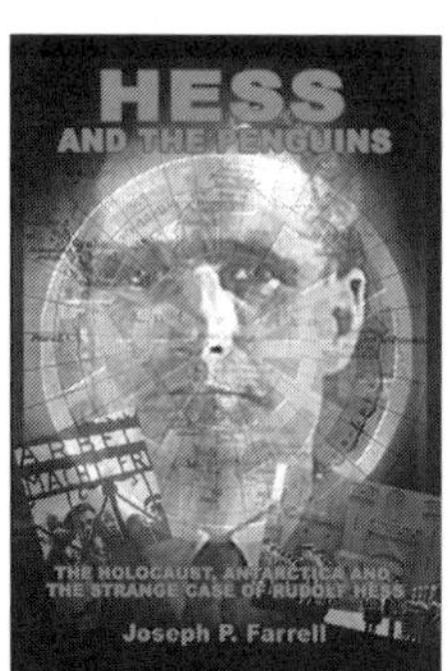

HESS AND THE PENGUINS
The Holocaust, Antarctica and the Strange Case of Rudolf Hess
By Joseph P. Farrell

Farrell looks at Hess' mission to make peace with Britain and get rid of Hitler—even a plot to fly Hitler to Britain for capture! How much did Göring and Hitler know of Rudolf Hess' subversive plot, and what happened to Hess? Why was a doppleganger put in Spandau Prison and then "suicided"? Did the British use an early form of mind control on Hess' double? John Foster Dulles of the OSS and CIA suspected as much. Farrell also uncovers the strange death of Admiral Richard Byrd's son in 1988, about the same time of the death of Hess.

288 Pages. 6x9 Paperback. Illustrated. $19.95. Code: HAPG

HIDDEN FINANCE, ROGUE NETWORKS & SECRET SORCERY
The Fascist International, 9/11, & Penetrated Operations
By Joseph P. Farrell

Farrell investigates the theory that there were not *two* levels to the 9/11 event, but *three*. He says that the twin towers were downed by the force of an exotic energy weapon, one similar to the Tesla energy weapon suggested by Dr. Judy Wood, and ties together the tangled web of missing money, secret technology and involvement of portions of the Saudi royal family. Farrell unravels the many layers behind the 9-11 attack, layers that include the Deutschebank, the Bush family, the German industrialist Carl Duisberg, Saudi Arabian princes and the energy weapons developed by Tesla before WWII.

296 Pages. 6x9 Paperback. Illustrated. $19.95. Code: HFRN

THRICE GREAT HERMETICA & THE JANUS AGE
By Joseph P. Farrell

What do the Fourth Crusade, the exploration of the New World, secret excavations of the Holy Land, and the pontificate of Innocent the Third all have in common? Answer: Venice and the Templars. What do they have in common with Jesus, Gottfried Leibniz, Sir Isaac Newton, Rene Descartes, and the Earl of Oxford? Answer: Egypt and a body of doctrine known as Hermeticism. The hidden role of Venice and Hermeticism reached far and wide, into the plays of Shakespeare (a.k.a. Edward DeVere, Earl of Oxford), into the quest of the three great mathematicians of the Early Enlightenment for a lost form of analysis, and back into the end of the classical era, to little known Egyptian influences at work during the time of Jesus.

354 Pages. 6x9 Paperback. Illustrated. $19.95. Code: TGHJ

THE CHILDREN OF MU
By James Churchward

According to Churchward, the lost Pacific continent of Mu was the site of the Garden of Eden and the home of 64,000,000 inhabitants known as the Naacals. Churchward tells the story of the colonial expansion of Mu and the influence of the highly developed Mu culture on the rest of the world. Her first colonies were in North America and the Orient, while other colonies had been started in India, Egypt and Yucatan. Chapters include: The Origin of Man; The Eastern Lines; Ancient North America; Stone Tablets from the Valley of Mexico; South America; Atlantis; Western Europe; The Greeks; Egypt; The Western Lines; India; Southern India; The Great Uighur Empire; Babylonia; Intimate Hours with the Rishi; more. Special photo section.

270 Pages. 6x9 Paperback. Illustrated. $19.95. Code: COMU

SECRET MARS: The Alien Connection
By Michael J. Craig

While scientists spend billions of dollars confirming that microbes live in the Martian soil, people sitting at home on their computers studying the Mars images are making far more astounding discoveries… they have found the possible archaeological remains of an extraterrestrial civilization. Hard to believe? Well, this challenging book invites you to take a look at the astounding pictures yourself and make up your own mind. *Secret Mars* presents over 160 incredible images taken by American and European spacecraft that reveal possible evidence of a civilization that once lived, and may still live, on the planet Mars… powerful evidence that scientists are ignoring! A visual and fascinating book!

352 Pages. 6x9 Paperback. Illustrated. $19.95. Code: SMAR

LBJ AND THE CONSPIRACY TO KILL KENNEDY
By Joseph P. Farrell

Farrell says that a coalescence of interests in the military industrial complex, the CIA, and Lyndon Baines Johnson's powerful and corrupt political machine in Texas led to the events culminating in the assassination of JFK. Chapters include: Oswald, the FBI, and the CIA: Hoover's Concern of a Second Oswald; Oswald and the Anti-Castro Cubans; The Mafia; Hoover, Johnson, and the Mob; The FBI, the Secret Service, Hoover, and Johnson; The CIA and "Murder Incorporated"; Ruby's Bizarre Behavior; The French Connection and Permindex; Big Oil; The Dead Witnesses: Guy Bannister, Jr., Mary Pinchot Meyer, Rose Cheramie, Dorothy Killgallen, Congressman Hale Boggs; LBJ and the Planning of the Texas Trip; LBJ: A Study in Character, Connections, and Cabals; LBJ and the Aftermath: Accessory After the Fact; The Requirements of Coups D'État; more.

342 Pages. 6x9 Paperback. $19.95 Code: LCKK

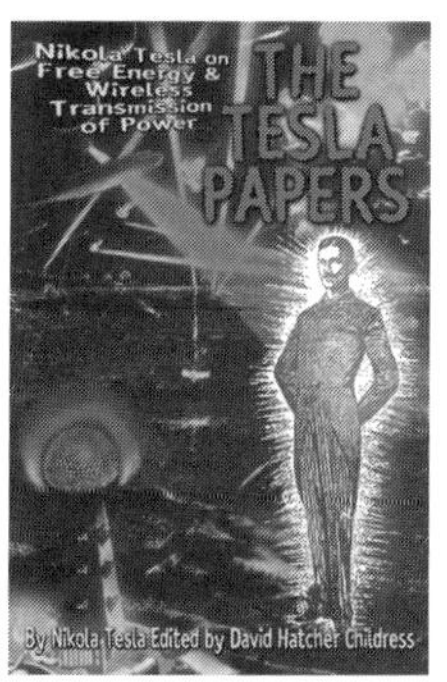

THE TESLA PAPERS
Nikola Tesla on Free Energy & Wireless Transmission of Power
by Nikola Tesla, edited by David Hatcher Childress

David Hatcher Childress takes us into the incredible world of Nikola Tesla and his amazing inventions. Tesla's fantastic vision of the future, including wireless power, anti-gravity, free energy and highly advanced solar power. Also included are some of the papers, patents and material collected on Tesla at the Colorado Springs Tesla Symposiums, including papers on: •The Secret History of Wireless Transmission •Tesla and the Magnifying Transmitter •Design and Construction of a Half-Wave Tesla Coil •Electrostatics: A Key to Free Energy •Progress in Zero-Point Energy Research •Electromagnetic Energy from Antennas to Atoms

325 PAGES. 8x10 PAPERBACK. ILLUSTRATED. $16.95. CODE: TTP

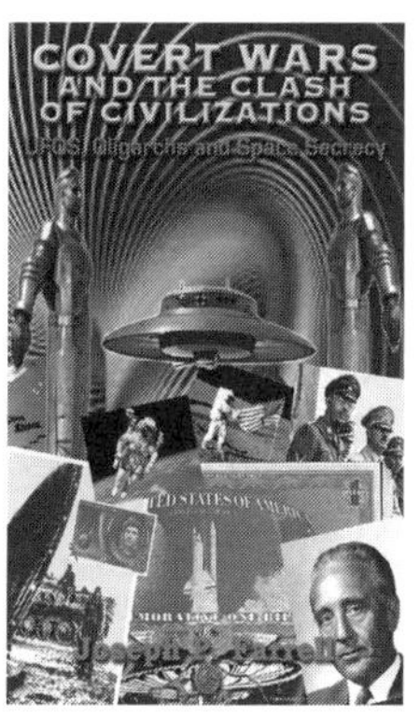

COVERT WARS & THE CLASH OF CIVILIZATIONS
UFOs, Oligarchs and Space Secrecy
By Joseph P. Farrell

Farrell's customary meticulous research and sharp analysis blow the lid off of a worldwide web of nefarious financial and technological control that very few people even suspect exists. He elaborates on the advanced technology that they took with them at the "end" of World War II and shows how the breakaway civilizations have created a huge system of hidden finance with the involvement of various banks and financial institutions around the world. He investigates the current space secrecy that involves UFOs, suppressed technologies and the hidden oligarchs who control planet earth for their own gain and profit.

358 Pages. 6x9 Paperback. Illustrated. $19.95. Code: CWCC

ANCIENT ALIENS ON THE MOON
By Mike Bara

What did NASA find in their explorations of the solar system that they may have kept from the general public? How ancient really are these ruins on the Moon? Using official NASA and Russian photos of the Moon, Bara looks at vast cityscapes and domes in the Sinus Medii region as well as glass domes in the Crisium region. Bara also takes a detailed look at the mission of Apollo 17 and the case that this was a salvage mission, primarily concerned with investigating an opening into a massive hexagonal ruin near the landing site. Chapters include: The History of Lunar Anomalies; The Early 20th Century; Sinus Medii; To the Moon Alice!; Mare Crisium; Yes, Virginia, We Really Went to the Moon; Apollo 17; more. Tons of photos of the Moon examined for possible structures and other anomalies.
248 Pages. 6x9 Paperback. Illustrated.. $19.95. Code: AAOM

ANCIENT ALIENS ON MARS
By Mike Bara

Bara brings us this lavishly illustrated volume on alien structures on Mars. Was there once a vast, technologically advanced civilization on Mars, and did it leave evidence of its existence behind for humans to find eons later? Did these advanced extraterrestrial visitors vanish in a solar system wide cataclysm of their own making, only to make their way to Earth and start anew? Was Mars once as lush and green as the Earth, and teeming with life? Chapters include: War of the Worlds; The Mars Tidal Model; The Death of Mars; Cydonia and the Face on Mars; The Monuments of Mars; The Search for Life on Mars; The True Colors of Mars and The Pathfinder Sphinx; more. Color section.
252 Pages. 6x9 Paperback. Illustrated. $19.95. Code: AMAR

ANCIENT ALIENS ON MARS II
By Mike Bara

Using data acquired from sophisticated new scientific instruments like the Mars Odyssey THEMIS infrared imager, Bara shows that the region of Cydonia overlays a vast underground city full of enormous structures and devices that may still be operating. He peels back the layers of mystery to show images of tunnel systems, temples and ruins, and exposes the sophisticated NASA conspiracy designed to hide them. Bara also tackles the enigma of Mars' hollowed out moon Phobos, and exposes evidence that it is artificial. Long-held myths about Mars, including claims that it is protected by a sophisticated UFO defense system, are examined. Data from the Mars rovers Spirit, Opportunity and Curiosity are examined; everything from fossilized plants to mechanical debris is exposed in images taken directly from NASA's own archives.
294 Pages. 6x9 Paperback. Illustrated. $19.95. Code: AAM2

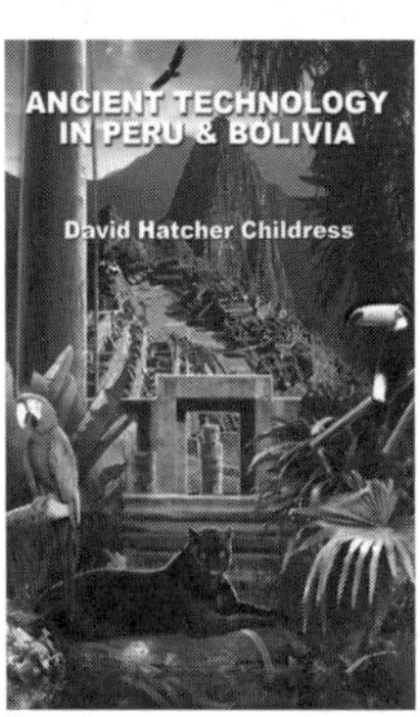

ANCIENT TECHNOLOGY IN PERU & BOLIVIA
By David Hatcher Childress

Childress speculates on the existence of a sunken city in Lake Titicaca and reveals new evidence that the Sumerians may have arrived in South America 4,000 years ago. He demonstrates that the use of "keystone cuts" with metal clamps poured into them to secure megalithic construction was an advanced technology used all over the world, from the Andes to Egypt, Greece and Southeast Asia. He maintains that only power tools could have made the intricate articulation and drill holes found in extremely hard granite and basalt blocks in Bolivia and Peru, and that the megalith builders had to have had advanced methods for moving and stacking gigantic blocks of stone, some weighing over 100 tons.
340 Pages. 6x9 Paperback. Illustrated.. $19.95 Code: ATP

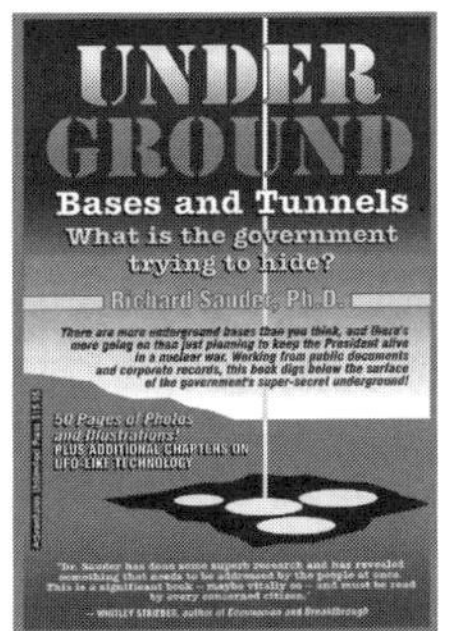

UNDERGROUND BASES & TUNNELS:
What is the Government Trying to Hide?
by Richard Sauder, Ph.D.

Working from government documents and corporate records, Sauder has compiled an impressive book that digs below the surface of the military's super-secret underground! Go behind the scenes into little-known corners of the public record and discover how corporate America has worked hand-in-glove with the Pentagon for decades, dreaming about, planning, and actually constructing, secret underground bases. This book includes chapters on the locations of the bases, the tunneling technology, various military designs for underground bases, abductions, needles & implants, military involvement in "alien" cattle mutilations, more. 50-page photo & map insert.

201 pages. 6x9 Paperback. Illustrated. $15.95. Code: UGB

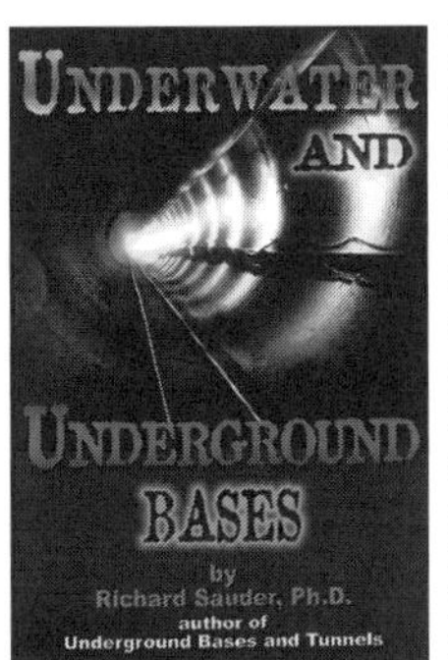

UNDERWATER & UNDERGROUND BASES
by Richard Sauder, Ph.D.

Dr. Sauder lays out the amazing evidence and government paper trail for the construction of huge, manned bases offshore, in mid-ocean, and deep beneath the sea floor! Official United States Navy documents, and other hard evidence, raise many questions about what really lies 20,000 leagues beneath the sea. Plus, breakthrough material reveals the existence of additional clandestine underground facilities as well as the surprising location of one of the CIA's own underground bases. Plus, information on tunneling and cutting-edge, high speed rail magnetic-levitation (MagLev) technology.

264 pages. 6x9 Paperback. Illustrated. $16.95. Code: UUB

AMERICAN CONSPIRACY FILES
The Stories We Were Never Told
By Peter Kross

Kross reports on conspiracies in the Revolutionary War, including those surrounding Benedict Arnold and Ben Franklin's son, William. He delves into the large conspiracy to kill President Lincoln and moves into our modern day with chapters on the deaths of JFK, RFK and MLK., the reasons behind the Oklahoma City bombing, the sordid plots of President Lyndon Johnson and more. Chapters on Edward Snowden; The Weather Underground; Patty Hearst; The Death of Mary Meyer; Marilyn Monroe; The Zimmerman Telegram; BCCI; Operation Northwinds; The Search for Nazi Gold; The Death of Frank Olsen; tons more. Over 50 chapters in all.

460 Pages. 6x9 Paperback. Illustrated. $19.95 Code: ACF

THE SECRET HISTORY OF THE UNITED STATES:
Conspiracies, Cobwebs and Lies
By Peter Kross

This book tells the stories of unexplained events in our history, as well as mysteries that have never been solved. The events covered in the book range from the American Revolution, the Civil War, World War II, the Cold War, the assassinations of the 1960s, the Iraq war and the events leading up to 9-11. Among the subjects covered are the following: the plot to oust FDR; Flight 19; Who killed JFK?; Nixon and the mob, Watergate and the CIA, Iran-Contra, and the intelligence failures that led up to 9-11. These stories are fascinating accounts of the underside of our history that will amaze the reader.

438 Pages. 6x9 Paperback. Illustrated. $19.95. Code: SHUS

ORDER FORM

10% Discount When You Order 3 or More Items

One Adventure Place
P.O. Box 74
Kempton, Illinois 60946
United States of America
Tel.: 815-253-6390 • Fax: 815-253-6300
Email: auphq@frontiernet.net
http://www.adventuresunlimitedpress.com

ORDERING INSTRUCTIONS

- ✓ Remit by USD$ Check, Money Order or Credit Card
- ✓ Visa, Master Card, Discover & AmEx Accepted
- ✓ Paypal Payments Can Be Made To: info@wexclub.com
- ✓ Prices May Change Without Notice
- ✓ 10% Discount for 3 or More Items

SHIPPING CHARGES

United States

- ✓ Postal Book Rate { $4.50 First Item; 50¢ Each Additional Item }
- ✓ POSTAL BOOK RATE Cannot Be Tracked! Not responsible for non-delivery.
- ✓ Priority Mail { $7.00 First Item; $2.00 Each Additional Item }
- ✓ UPS { $9.00 First Item (Minimum 5 Books); $1.50 Each Additional Item }

NOTE: UPS Delivery Available to Mainland USA Only

Canada

- ✓ Postal Air Mail { $19.00 First Item; $3.00 Each Additional Item }
- ✓ Personal Checks or Bank Drafts MUST BE US$ and Drawn on a US Bank
- ✓ Canadian Postal Money Orders OK
- ✓ Payment MUST BE US$

All Other Countries

- ✓ Sorry, No Surface Delivery!
- ✓ Postal Air Mail { $19.00 First Item; $7.00 Each Additional Item }
- ✓ Checks and Money Orders MUST BE US$ and Drawn on a US Bank or branch.
- ✓ Paypal Payments Can Be Made in US$ To: info@wexclub.com

SPECIAL NOTES

- ✓ RETAILERS: Standard Discounts Available
- ✓ BACKORDERS: We Backorder all Out-of-Stock Items Unless Otherwise Requested
- ✓ PRO FORMA INVOICES: Available on Request
- ✓ DVD Return Policy: Replace defective DVDs only

ORDER ONLINE AT: www.adventuresunlimitedpress.com

10% Discount When You Order 3 or More Items!

Please check: ✓

☐ This is my first order ☐ I have ordered before

Name

Address

City

State/Province | Postal Code

Country

Phone: Day | Evening

Fax | Email

Item Code	Item Description	Qty	Total

Please check: ✓

☐ Postal-Surface

☐ Postal-Air Mail (Priority in USA)

☐ UPS (Mainland USA only)

☐ Visa/MasterCard/Discover/American Express

Subtotal ▶

Less Discount-10% for 3 or more items ▶

Balance ▶

Illinois Residents 6.25% Sales Tax ▶

Previous Credit ▶

Shipping ▶

Total (check/MO in USD$ only) ▶

Card Number:

Expiration Date: | Security Code:

✓ SEND A CATALOG TO A FRIEND: